MATHEMATICAL MODELING IN DIFFRACTION THEORY

MATHEMATICAL MODELING IN DIFFRACTION THEORY

Based on A Priori Information on the Analytic Properties of the Solution

ALEXANDER G. KYURKCHAN

NADEZHDA I. SMIRNOVA

Moscow Technical University of Communications and Informatics, Moscow, Russia

ELSEVIER

AMSTERDAM • BOSTON • HEIDELBERG • LONDON • NEW YORK • OXFORD
PARIS • SAN DIEGO • SAN FRANCISCO • SINGAPORE • SYDNEY • TOKYO

Elsevier
Radarweg 29, PO Box 211, 1000 AE Amsterdam, Netherlands
The Boulevard, Langford Lane, Kidlington, Oxford OX5 1GB, UK
225 Wyman Street, Waltham, MA 02451, USA

Notices
Knowledge and best practice in this field are constantly changing. As new research and
experience broaden our understanding, changes in research methods, professional
practices, or medical treatment may become necessary.

Practitioners and researchers must always rely on their own experience and knowledge in
evaluating and using any information, methods, compounds, or experiments described
herein. In using such information or methods they should be mindful of their own safety
and the safety of others, including parties for whom they have a professional responsibility.

To the fullest extent of the law, neither the Publisher nor the authors, contributors,
or editors, assume any liability for any injury and/or damage to persons or property
as a matter of products liability, negligence or otherwise, or from any use or operation
of any methods, products, instructions, or ideas contained in the material herein.

Library of Congress Cataloging-in-Publication Data
A catalog record for this book is available from the Library of Congress

British Library Cataloguing in Publication Data
A catalogue record for this book is available from the British Library

ISBN: 978-0-12-803728-7

For information on all Elsevier publications
visit our website at http://store.elsevier.com/

Working together
to grow libraries in
developing countries

www.elsevier.com • www.bookaid.org

CONTENTS

INTRODUCTION

In many fields of contemporary science, from astrophysics to biology, scientists are required to find effective solutions for wave diffraction and scattering problems. This is, in particular, confirmed by the programs of regular conferences such as "Days on Diffraction," "Mathematical Methods in Electromagnetic Theory," "Electromagnetic and Light Scattering," "PIERS," etc.

The use of *a priori* information permits a significant increase in the efficiency of algorithms for solving such problems, and in several cases, the problem can principally only be solved by using such information. Here, we consider several examples to illustrate this.

1. Diffraction at thin screens. In problems of wave diffraction at thin screens, there is a complexity related to the fact that the current $I(x)$ on the screen has a singularity near the screen edge. It follows from the Meixner condition [1] that the current component parallel to the screen edge has a singularity of the form $\rho^{-1/2}$, where ρ is the distance to the screen edge. Thus, for example, when solving the problem of diffraction at an infinitely thin band of width $2a$ (in the case of E-polarization), it is expedient to represent the desired current as

$$I(x) = \frac{1}{\sqrt{a^2 - x^2}} J(x),$$

where $J(x)$ is now a smooth function.

If the boundary-value problem reduces to an integral equation, then the use of such a representation allows one to simplify the corresponding algorithm to solving this equation.

2. Diffraction at a periodic grating. We now consider the problem of diffraction of a plane wave at a periodic grating consisting of N elements. If there are infinitely many elements, then the problem reduces to a single period of the grating. If N is not large, then it is necessary to solve the problem of diffraction at a group of bodies. The situation is most difficult when N is finite but rather large. In this case, it is very difficult to solve the problem of diffraction at a group of N bodies because of the extremely large volume of calculations. Here it is appropriate to use *a priori* information stating that the distribution of fields or currents on $M < N$ central elements of the grating is approximately the same as the distribution on an infinite periodic grating [2]. Thus, the problem reduces to determining the field distribution on $(N - M) + 1$ elements of the grating, namely, on $N - M$ edge elements and one central element, which already allows the problem to be solved numerically.

3. Wave scattering at a body whose dimensions are much more than the wavelength. In this case, the use of the standard methods, for example, using integral equations, is not efficient because algebraization results in a system of large-sized algebraic equations. Here, it is expedient to use the hybrid method, where the current on a greater part of the scatterer is assumed to be equal to the current of geometric optics, i.e., the unknown current is equal to the current on the corresponding tangent plane on the illuminated part of the scatterer, and to zero on the shaded part [3]. The current remains unknown only near the "light-shadow" boundary, which permits a significant decrease in the dimensions of the corresponding algebraic system.

In the second and third examples, we discuss the use of *a priori* information of physical character, which was obtained by solving special problems. Already in the first example, we discuss the use of information about the analytic properties of the solution, which is directly contained in the boundary condition (the Meixner condition). This condition clearly contains the required information. From this standpoint, the problems of wave diffraction at smooth bodies, where the boundary conditions do not give any explicit information about the analytic properties of the solution, are much more complicated. We consider the diffraction problem for a monochromatic wave with the time-dependence $f(t) = e^{i\omega t}$ (here ω is the cyclic frequency of oscillations) propagating in a homogeneous isotropic medium with relative dielectric permittivity ε and relative magnetic permeability μ. The mathematical model of such a problem is the system of Maxwell equations for the vectors of electromagnetic field strength with boundary conditions or, in the scalar approximation, the external boundary-value problem for the Helmholtz equation [1,4,5]. As is known in this case (e.g., see [5,6]), the solutions of these equations are real analytic functions that we call *wave fields*. The direct diffraction problem consists of determining the secondary wave field arising when the known primary field meets a certain obstacle, i.e., a scatterer. Throughout the monograph, we deal only with external problems of diffraction. The solutions of such problems are defined as everywhere outside the scatterer, i.e., outside a certain domain D with boundary S. Outside D, the wave fields vanish at infinity according to the Sommerfeld radiation condition [1,5]. This means that the wave field must have singularities in the domain D (or on its boundary), because otherwise (as will be shown below) the field would be zero.

We consider another example that illustrates the importance of such information.

Assume that we must solve the problem of diffraction of a primary wave field $U^0(\vec{r})$ at a compact scatterer bound by the surface S. When solving this problem, one can use the following representation for the field $U^1(\vec{r})$ scattered at the body [4,5]

$$U^1(r, \theta, \varphi) = \sum_{n=0}^{\infty} \sum_{m=-n}^{n} a_{nm}(-i)^{n+1} h_n^{(2)}(kr) P_n^m(\cos\theta) e^{im\varphi}. \qquad (1)$$

A similar representation in the form of a series in spherical harmonics also holds in the case of the vector [7]. Substituting this expansion for $U^1(\vec{r})$ in the boundary condition, we can determine the coefficients a_{nm} and hence solve the boundary-value problem. However, this can only be performed if the series written for $U^1(\vec{r})$ converges to the scatterer boundary S. This assumption is known in the literature as the Rayleigh hypothesis [8,9] and the bodies satisfying this hypothesis are called *Rayleigh bodies*. The series in spherical harmonics for the scattered field $U^1(\vec{r})$ is a power series in its infinite residuals (see below), and hence converges only outside the sphere containing all singularities of the function $U^1(\vec{r})$. Thus, only the bodies that are completely contained inside this sphere S are Rayleigh bodies. To know whether the body is Rayleigh or not is obviously of principal importance when solving the diffraction problem by the method described above. The fact that this information can be obtained *a priori* is also important [9,10].

What are these singularities of the function $U^1(\vec{r})$? We consider a simple example. Assume that a source of light is located in front of a plane mirror S. When looking at a mirror, we see the source of light at a point behind the mirror that is symmetric with respect to S (see Fig. 1). We see

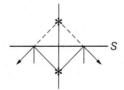

Figure 1 Image in a flat mirror.

this imaginary source (image) as the straight continuation of the rays reflected from the mirror. The source image is precisely the result of analytic continuation of the field reflected from the mirror into the region behind the mirror.

We perceive the obtained image as an additional source of light on the other side of the mirror as if we can see "though the mirror." Moreover, if we begin to move the source of light away from the mirror, then its image "behind the mirror" also moves symmetrically as far as possible.

The situation is quite different in the case of reflection from a nonplane surface. We now imagine that we begin to bend a plane mirror by lifting its "edges." The space behind the mirror begins to "contract" forming a "fold." A part of the image or even the entire image of the object may then disappear (this effect is familiar to everybody who has looked at their image in halls of mirrors, recall the laughter rooms). This effect arises due to formation of "folds" in the hypothetical media behind the mirror, i.e., of regions containing two (or even more) images simultaneously. In this case, the disappearing part of the image is "hidden in the fold". A more detailed qualitative discussion of this effect is available (see [10]).

In the case of problems of diffraction at compact obstacles, the wave field (scattered at an obstacle) must also have singularities. Obviously, these singularities must lie outside the domain, where we seek the diffraction field (in the so-called nonphysical region). Thus, we speak of singularities of the analytic continuation of the wave field beyond the original domain of its definition.

In Chapter 1, we discus several methods required to localize the singularities of the analytic continuation of wave fields in detail. In the subsequent chapters, we describe the methods for solving problems of the diffraction theory that are significantly based on the information about the singularities of the analytic continuation of the wave field. Let us briefly list the required methods.

One of the most widely used methods for solving problems of diffraction is the reduction of the corresponding boundary-value problem to integral equations [5,11,12]. The popularity of approaches based on the techniques of (current) integral equations can, in particular, be explained by the combination of their universality and high computing speed of such methods. However, the kernels of these integral equations are singular functions (i.e., they have singularities at the points of coincidence of the values of their arguments). This fact leads to certain technical difficulties in numerical computations.

Different modifications of integral equation methods based on the idea of separating the surfaces, namely, the surface where the observation point is chosen and the surface that is the current support, are being used more frequently. Such a separation of surfaces allows one to eliminate the singularity in the kernel of the integral equation without any additional computations and thus to simplify the calculations and reduce their volume. This idea underlies the continued boundary condition method (CBCM) [13,14], the method of auxiliary currents (MAC) [15,16] and its special case, the method of discrete (auxiliary) sources (MDS) [17,18], the null field method (NFM) [5,19] and its popular implementation, the T-matrix method (TMM) [19,20]. Because the terminology is rather inaccurate, which we discuss later, CBCM is sometimes confused with NFM, while TMM and NFM are considered as the same methods.

In CBCM, the surface where we choose an observation point, which we denote by S_δ, lies outside the scatterer at a certain sufficiently small distance δ from its boundary S, which is the carrier of the (auxiliary) current and is the domain of integration. Since the wave field is analytic, the boundary condition is approximately satisfied on the surface S_δ and, as a result, the diffraction problem is solved in an approximate setting. The main advantages of CBCM are its universality and simplicity [14,21]. The main drawback of CBCM is a lower rate of convergence compared, for example, with MAC. MAC and NFM (as was shown later [22,23]) can be used only in the case of scatterers with an analytic boundary that is a closed surface. However, if these methods are applicable, they, in contrast to CBCM, permit solving the diffraction problem in the exact statement and ensure a very high rate of convergence of numerical algorithms used to solve the corresponding integral equations.

In MAC, the surface Σ, i.e., the carrier of the auxiliary current, is located inside the scatterer surface S, where the boundary condition is satisfied. According to the assumptions of the existence theorem (see below), the surface Σ must surround the set $\bar{A}$ of singularities of the analytic continuation of the scattered field [9,10,16]. In NFM, the carrier of the sought current is the scatterer surface and the observation point lies on a certain surface Σ inside the scatterer, where the null field condition is satisfied. As was shown in the late 2000s [22,23], the interior of the surface Σ in NFM must also contain the set $\bar{A}$.

MDS and TMM were the first methods that appeared almost simultaneously [17,19]. MDS is a special case of MAC, where the continuous distribution of an auxiliary current is replaced by a linear combination of discrete sources. Later, because there were no criteria for optimal choice of the surface (the source support), some other more difficult implementations of MDS were developed, for example, the methods based on the use of multipoles and complex sources [24,25]. In the literature, one can find various methods for constructing the surface Σ, and the methods, where Σ is a surface similar to or equidistant from the scatter surface S, are used most widely. In 2001, it was shown that the construction of the surface Σ in MDS by using the techniques of analytic deformation of the surface S ensures that the assumptions of the existence theorem are satisfied for the corresponding integral equation and the rate of convergence of the computational algorithm is maximally high [26]. Such a version of the method, where the surface Σ is constructed by analytic deformation of the surface S, was called a modified method of discrete sources (MMDS). Evidently, the results obtained in that paper also hold for MAC.

The importance of *a priori* analytic information became apparent in the early 2000s and, as an analysis of the literature shows, such information is actively used by only a narrow group of researchers. This can be explained by the fact that there are no monographs specifically on this subject. At present, there is only one monograph (in Russian) [9], where Chapter 2 presents the basic ideas of how the information about the analytic properties of wave fields can be used to solve the diffraction problems, and the collective monograph [25], where Chapter 5 mainly deals with only one aspect of this usage. In the two monographs devoted to the methods of mathematical modeling in the theory of diffraction, there is little focus on the problems of application of *a priori* information about the analytic properties of the solution [27,28].

CHAPTER 1

Analytic Properties of Wave Fields

The series in spherical harmonics (Eqn. 1) considered in the "Introduction" is only one of the methods for obtaining analytic representations of wave fields. The diffraction problems can also be solved by different methods for such representations. The basic (and most widely used) methods for analytic representations are considered below.

1.1 DERIVATION OF BASIC ANALYTIC REPRESENTATIONS OF WAVE FIELDS

1.1.1 Representation of Fields by Wave Potential

To represent the field in terms of (wave) potentials is one of the most widely used methods for solving the boundary-value problems of diffraction theory.

We consider the problem of wave diffraction at a compact scatterer occupying a spatial domain D with boundary S. Let $\bar{A}$ be the set of singularities of the diffraction field $U^1(\vec{r})$, that is, of a function that everywhere outside S satisfies the homogeneous Helmholtz equation,

$$\Delta U^1(\vec{r}) + k^2 U^1(\vec{r}) = 0, \tag{1.1}$$

where $k = 2\pi/\lambda$ is the wave number and λ is the wavelength, the Sommerfeld condition at infinity [1], and a certain boundary condition on S.

Let Σ be a surface (a contour in the two-dimensional case) surrounding $\bar{A}$. Then, we can show that everywhere outside Σ, the field $U^1(\vec{r})$ can be represented by the integral,

$$U^1(\vec{r}) = \int_{\Sigma} \left\{ U^1(\vec{r}') \frac{\partial G_0(\vec{r}; \vec{r}')}{\partial n'} - \frac{\partial U^1(\vec{r}')}{\partial n'} G_0(\vec{r}; \vec{r}') \right\} d\sigma', \tag{1.2}$$

where $G_0(\vec{r}; \vec{r}') = \frac{1}{4\pi} \frac{\exp(-ik|\vec{r}-\vec{r}'|)}{|\vec{r}-\vec{r}'|}$ (in the three-dimensional case) or $G_0(\vec{r}; \vec{r}') = \frac{1}{4i} H_0^{(2)}(k|\vec{r}-\vec{r}'|)$ (in the two-dimensional case) is the fundamental solution of the Helmholtz equation (the Green function of the free space) and $\frac{\partial}{\partial n'}$ denotes the differentiation in the direction of the outer normal on Σ.

Mathematical Modeling in Diffraction Theory
http://dx.doi.org/10.1016/B978-0-12-803728-7.00001-4

Formula (1.2) can be obtained, for example, as follows. Along with (1.1), we consider the equation for the function $G_0(\vec{r}; \vec{r}')$,

$$\Delta G_0(\vec{r}; \vec{r}') + k^2 G_0(\vec{r}; \vec{r}') = -\delta(\vec{r}, \vec{r}'), \qquad (1.3)$$

where $\delta(\vec{r}, \vec{r}')$ is the Dirac delta function. Further, we multiply Equation (1.1) by $G_0(\vec{r}; \vec{r}')$ and Equation (1.3) by $U^1(\vec{r})$, subtract the second relation from the first, and integrate the result over the entire space outside the surface Σ. Then we obtain (1.2) applying the Ostrogradsky theorem and using the condition at infinity

$$\lim_{r \to \infty} r^{(n-1)/2} \left(\frac{\partial U^1(\vec{r})}{\partial r} + ik U^1(\vec{r}) \right) = 0, \quad n = 2, 3. \qquad (1.4)$$

Using the second Green formula for the domain Ω between the surfaces S (the scatterer boundary) and Σ, we obtain

$$U^1(\vec{r}) = \int_{S+\Sigma} \left\{ U^1(\vec{r}') \frac{\partial G_0(\vec{r}; \vec{r}')}{\partial n'} - \frac{\partial U^1(\vec{r}')}{\partial n'} G_0(\vec{r}; \vec{r}') \right\} ds'. \qquad (1.5)$$

Thus, it follows from (1.2) and (1.5) that

$$\int_S \left\{ U^1(\vec{r}') \frac{\partial G_0(\vec{r}; \vec{r}')}{\partial n'} - \frac{\partial U^1(\vec{r}')}{\partial n'} G_0(\vec{r}; \vec{r}') \right\} ds' = 0, \qquad (1.6)$$

for all points $M(\vec{r}) \in \Omega$. The integral on the left-hand side in (1.6) is an analytic function everywhere inside D, and, because this integral is zero in the subdomain Ω inside D, it follows that relation (1.6) holds for all points $M(\vec{r}) \in D$.

Thus, we have the relation [5]

$$\int_S \left\{ U^1(\vec{r}') \frac{\partial G_0(\vec{r}; \vec{r}')}{\partial n'} - \frac{\partial U^1(\vec{r}')}{\partial n'} G_0(\vec{r}; \vec{r}') \right\}$$

$$ds' = \begin{cases} U^1(\vec{r}), & M(\vec{r}) \in R^n \backslash \bar{D}, \\ 0, & M(\vec{r}) \in D, \end{cases} \qquad (1.7)$$

where $n = 2, 3$.

Similarly, for the internal wave field $U^i(\vec{r})$ that satisfies the equation

$$\Delta U^i(\vec{r}) + k^2 U^i(\vec{r}) = 0$$

in the domain D, we have [5]

$$\int_S \left\{ U^i(\vec{r}'') \frac{\partial G_0(\vec{r};\vec{r}'')}{\partial n'} - \frac{\partial U^i(\vec{r}'')}{\partial n'} G_0(\vec{r};\vec{r}'') \right\}$$

$$ds' = \begin{cases} 0, & M(\vec{r}) \in R^n \backslash \bar{D}, \\ -U^i(\vec{r}), & M(\vec{r}) \in D. \end{cases} \tag{1.8}$$

Since the initial (incident) field $U^0(\vec{r})$ satisfies the relation

$$\int_S \left\{ U^0(\vec{r}'') \frac{\partial G_0(\vec{r};\vec{r}'')}{\partial n'} - \frac{\partial U^0(\vec{r}'')}{\partial n'} G_0(\vec{r};\vec{r}'') \right\} ds' = -U^0(\vec{r}),$$

everywhere inside S, we can rewrite (1.7) as

$$\int_S \left\{ U(\vec{r}'') \frac{\partial G_0(\vec{r};\vec{r}'')}{\partial n'} - \frac{\partial U(\vec{r}'')}{\partial n'} G_0(\vec{r};\vec{r}'') \right\}$$

$$ds' = \begin{cases} U^1(\vec{r}), & M(\vec{r}) \in R^n \backslash \bar{D}, \\ -U^0(\vec{r}), & M(\vec{r}) \in D, \end{cases} \tag{1.9}$$

where $U = U^0 + U^1$ is the complete wave field.

In the literature, the obtained representations (1.7)–(1.9) are usually called wave potentials [5].

Relation (1.9) can be more convenient, because it can be simplified with the boundary condition on S taken into account. For example, in the case of a boundary condition of the form

$$\left. \left(\alpha U + \beta \frac{\partial U}{\partial n} \right) \right|_S = 0, \tag{1.10}$$

where α, β are some given coefficients and $\dfrac{\partial}{\partial n}$ denotes differentiation in the direction of the outer normal to S, formula (1.9) can be rewritten as

$$U^0(\vec{r}) - \int_S \frac{\partial U(\vec{r}'')}{\partial n'} \left\{ \frac{\beta}{\alpha} \frac{\partial G_0(\vec{r};\vec{r}'')}{\partial n'} + G_0(\vec{r};\vec{r}'') \right\} ds' = 0, \quad M(\vec{r}) \in D.$$

Because

$$\int_{\Sigma}\left\{U^0(\vec{r}')\frac{\partial G_0(\vec{r};\vec{r}')}{\partial n'} - \frac{\partial U^0(\vec{r}')}{\partial n'}G_0(\vec{r};\vec{r}')\right\}$$

$$\mathrm{d}s' = 0, \quad M(\vec{r}) \in R^n\backslash\overline{\Omega}_{\Sigma}, \tag{1.11}$$

where Ω_{Σ} is a domain inside Σ, integral (1.2) can be rewritten as

$$U^1(\vec{r}) = \int_{\Sigma}\left\{U(\vec{r}')\frac{\partial G_0(\vec{r};\vec{r}')}{\partial n'} - \frac{\partial U(\vec{r}')}{\partial n'}G_0(\vec{r};\vec{r}')\right\}\mathrm{d}\sigma'. \tag{1.2a}$$

Relations of the forms given in Equations (1.7) and (1.9) and their analogs in the vector case (see below) are widely used in the classical method of current integral equations, as well as in the continued boundary condition method (CBCM) and the null field method (NFM) [5,13,14,19–23,28,29] (see Chapters 3 and 4). Representations (1.2), (1.2a), and their analogs for the vector diffraction problems underlie the methods of auxiliary currents and discrete sources ([9,16,25,30] and Chapter 2).

The scattered field representations of Equation (1.9) are also used in the problems of wave scattering at periodic boundaries. For example, in the two–dimensional case, the function $G_0(\vec{r};\vec{r}')$ in formula (1.9) has the form [9, Chapter 2]

$$G_0(\vec{r};r') = \frac{1}{2bi}\sum_{n=-\infty}^{\infty}\frac{\exp[-iw_n(x-x') - iv_n|y-y'|]}{v_n}, \tag{1.12}$$

where b is the period of the surface described by the equation $y = h(x) = h(x+b)$, $w_n = \frac{2\pi}{b}n + k\sin\theta$, $v_n = \sqrt{k^2 - w_n^2}$, and $v_n = -i\sqrt{w_n^2 - k^2}$ for $|w_n| > k$; here θ is the angle of incidence of the initial plane wave.

Now let us study the vector case. We consider the system of Maxwell equations [31,32]

$$\begin{cases} \nabla\times\vec{E} = -ik\zeta\vec{H} - \vec{J}^m, \\ \nabla\times\vec{H} = \dfrac{ik}{\zeta}\vec{E} + \vec{J}^e, \\ \nabla\vec{E} = 0, \ \nabla\vec{H} = 0, \end{cases} \tag{1.13}$$

where $\vec{E}, \vec{H}$ are the vectors of electric and magnetic field intensities, $k = \omega/c = 2\pi/\lambda$ is the wave number, ω is the angular frequency, c is the light speed, λ is the wavelength, $\zeta = \sqrt{\mu/\varepsilon}$ is the medium wave resistance, μ, ε are the magnetic permeability and dielectric permittivity of the medium, and $\vec{J}^e$, $\vec{J}^m$ are the densities of the electric and magnetic currents generating the fields $\vec{E}$, $\vec{H}$.

We solve the system of Equation (1.13) using the Fourier transform method. We set

$$\vec{E} = \frac{1}{8\pi^3} \iint \int_{-\infty}^{\infty} \vec{f}_E(\vec{w}) e^{-i\vec{w}\vec{r}} \, dw, \quad \vec{H} = \frac{1}{8\pi^3} \iint \int_{-\infty}^{\infty} \vec{f}_H(\vec{w}) e^{-i\vec{w}\vec{r}} \, dw, \quad (1.14)$$

where $\vec{r} = \{r_1, r_2, r_3\}$ is the radius vector of the observation point, $\vec{w} = \{w_1, w_2, w_3\}$, and $dw = dw_1 \, dw_2 \, dw_3$.

We use the vector identities $\nabla \times \vec{a}\,\psi = -(\vec{a} \times \nabla \psi)$, $\nabla \psi = -i\vec{w}\,\psi$, where $\psi = \exp(-i\vec{w}\,\vec{r})$ and $\vec{a}$ is a vector independent of $\vec{r}$, to obtain the equations

$$\frac{1}{8\pi^3} \iint \int_{-\infty}^{\infty} \left\{ i\left(\vec{f}_H \times \vec{w}\right) - i\omega\varepsilon \vec{f}_E \right\} e^{-i\vec{w}\vec{r}} \, dw = \vec{J}^e,$$

$$\frac{1}{8\pi^3} \iint \int_{-\infty}^{\infty} \left\{ i\left(\vec{f}_E \times \vec{w}\right) + i\omega\mu \vec{f}_H \right\} e^{-i\vec{w}\vec{r}} \, dw = -\vec{J}^m,$$

from (1.13) and (1.14), which imply

$$\left(\vec{f}_H \times \vec{w}\right) - \omega\varepsilon \vec{f}_E = -i\vec{f}^e \equiv -i \int_{V^e} \vec{J}^e e^{i\vec{w}\vec{r}} \, dv,$$

$$\left(\vec{f}_E \times \vec{w}\right) + \omega\mu \vec{f}_E = i\vec{f}^m \equiv i \int_{V^m} \vec{J}^m e^{i\vec{w}\vec{r}} \, dv,$$

where $V^e \equiv \text{supp } \vec{J}^e$ and $V^m \equiv \text{supp } \vec{J}^m$.

We substitute one equation into the other to obtain

$$\left(k^2 - w^2\right)\vec{f}_E + \vec{w}\left(\vec{w} \cdot \vec{f}_E\right) = i\omega\mu \vec{f}^e - i\left(\vec{w} \times \vec{f}^m\right),$$

$$\left(k^2 - w^2\right)\vec{f}_H + \vec{w}\left(\vec{w} \cdot \vec{f}_H\right) = i\omega\varepsilon \vec{f}^m + i\left(\vec{w} \times \vec{f}^e\right).$$

These equations allow us to determine

$$\left(k^2 - w^2\right)\vec{f}_E = i\omega\mu \vec{f}^e - \frac{i\omega\mu}{k^2} \vec{w}\left(\vec{w} \cdot \vec{f}^e\right) - i\left(\vec{w} \times \vec{f}^m\right) \equiv \vec{p}_E,$$

$$\left(k^2 - w^2\right)\vec{f}_H = i\omega\varepsilon \vec{f}^m - \frac{i\omega\varepsilon}{k^2} \vec{w}\left(\vec{w} \cdot \vec{f}^m\right) + i\left(\vec{w} \times \vec{f}^e\right) \equiv \vec{p}_H. \quad (1.15)$$

Now it follows from (1.14) and (1.15) that

$$\vec{E} = \frac{1}{8\pi^3} \int \int_{-\infty}^{\infty} \int \frac{\vec{P}_E(\vec{w})}{k^2 - w^2} e^{-i\vec{w}\vec{r}} \, dw,$$

$$\vec{H} = \frac{1}{8\pi^3} \int \int_{-\infty}^{\infty} \int \frac{\vec{P}_H(\vec{w})}{k^2 - w^2} e^{-i\vec{w}\vec{r}} \, dw.$$

We close the contour of integration over the variable w_3 in the upper and lower half-planes according to the Jordan lemma, take the contributions of poles at the points $w_3 = \pm\sqrt{k^2 - w_1^2 - w_2^2}$ into account (for $w_1^2 + w_2^2 > k^2$, we choose the root branch such that $\sqrt{k^2 - w_1^2 - w_2^2} = -i\sqrt{w_1^2 + w_2^2 - k^2}$), and obtain

$$\vec{E} = \frac{i}{8\pi^2} \int_{-\infty}^{\infty} \int_{-\infty}^{\infty} \vec{P}_E\left(w_1, w_2, \sqrt{k^2 - w_1^2 - w_2^2}\right) \frac{e^{-i\left(w_1 x + w_2 y \pm z\sqrt{k^2 - w_1^2 - w_2^2}\right)}}{\sqrt{k^2 - w_1^2 - w_2^2}} dw_1 \, dw_2,$$

$$\vec{H} = \frac{i}{8\pi^2} \int_{-\infty}^{\infty} \int_{-\infty}^{\infty} \vec{P}_H\left(w_1, w_2, \sqrt{k^2 - w_1^2 - w_2^2}\right) \frac{e^{-i\left(w_1 x + w_2 y \pm z\sqrt{k^2 - w_1^2 - w_2^2}\right)}}{\sqrt{k^2 - w_1^2 - w_2^2}} dw_1 \, dw_2.$$

In these formulas, the upper sign is taken for $z > z'_{\max} \equiv a$, and the lower sign, for $z < z'_{\min} \equiv -b$ (here $z'_{\max}$ and $z'_{\min}$ are the maximal and minimal values of the variable z' on the set $V = V^e \cup V^m$).

We change the variables as

$$w_1 = k \sin\alpha \cos\beta, \quad w_2 = k \sin\alpha \sin\beta.$$

Then we obtain

$$\sqrt{k^2 - w_1^2 - w_2^2} = k \cos\alpha, \, dw_1 \, dw_2 = k^2 \sin\alpha \cos\alpha \, d\alpha \, d\beta.$$

As a result, we have

$$\vec{E} = \frac{1}{2\pi i} \int_0^{2\pi} \int_0^{\pi/2 + i\infty} \vec{E}_0(\alpha, \beta) e^{-ik\left(\vec{k}_0 \vec{r}\right)} \sin\alpha \, d\alpha \, d\beta,$$

$$\vec{H} = \frac{1}{2\pi i} \int_0^{2\pi} \int_0^{\pi/2 + i\infty} \vec{H}_0(\alpha, \beta) e^{-ik\left(\vec{k}_0 \vec{r}\right)} \sin\alpha \, d\alpha \, d\beta. \qquad (1.16)$$

The obtained relations (1.16) are the wave field representations by the Sommerfeld-Weyl integrals [33] or by the integrals of plane waves (see below).

In these formulas, we have

$$\vec{E}_0(\alpha, \beta) = \frac{k^2}{4\pi i} \int_V \left\{ \zeta \left[\vec{J}^e(\vec{r}') - \vec{k}_0 \left(\vec{k}_0 \cdot \vec{J}^e(\vec{r}') \right) \right] \right.$$

$$\left. - \left(\vec{k}_0 \times \vec{J}^m(\vec{r}') \right) \right\} e^{ik\left(\vec{k}_0 \cdot \vec{r}' \right)} dv', \qquad (1.17)$$

$$\vec{H}_0(\alpha, \beta) = \frac{k^2}{4\pi i} \int_V \left\{ \frac{1}{\zeta} \left[\vec{J}^m(\vec{r}') - \vec{k}_0 \left(\vec{k}_0 \cdot \vec{J}^m(\vec{r}') \right) \right] \right.$$

$$\left. + \left(\vec{k}_0 \times \vec{J}^e(\vec{r}') \right) \right\} e^{ik\left(\vec{k}_0 \cdot \vec{r}' \right)} dv', \qquad (1.18)$$

where

$$\vec{k}_0 = \{ \sin\alpha \, \cos\beta, \sin\alpha \, \sin\beta, \cos\alpha \}, \quad \vec{r} = \{ r \sin\theta \, \cos\varphi, r \sin\theta \, \sin\varphi, r \cos\theta \}.$$

We substitute (1.17) and (1.18) into (1.16). Then, because

$$\nabla \times \vec{b} \, \phi = ik \left(\vec{b} \times \vec{k}_0 \right) \phi, \quad \left(\nabla \times \left(\nabla \times \vec{b} \phi \right) \right) = k^2 \left[\vec{b} - \vec{k}_0 \left(\vec{k}_0 \cdot \vec{b} \right) \right] \phi,$$

(where $\phi = \exp\left(-ik\vec{k}_0 \, \vec{r} \right)$ and $\vec{b}$ is a vector independent of $\vec{r}$), we have

$$\vec{E}(\vec{r}) = \int_V \left\{ \frac{\zeta}{ik} \left(\nabla \times \left(\nabla \times \vec{J}^e G_0 \right) \right) - \left(\nabla \times \vec{J}^m G_0 \right) \right\} dv',$$

$$\vec{H}(\vec{r}) = \int_V \left\{ \frac{1}{ik\zeta} \left(\nabla \times \left(\nabla \times \vec{J}^m G_0 \right) \right) + \left(\nabla \times \vec{J}^e G_0 \right) \right\} dv'. \qquad (1.19)$$

In these formulas,

$$G_0 = \frac{k}{8\pi^2 i} \int_0^{2\pi} \int_0^{\pi/2 + i\infty} e^{-ik\vec{k}_0 \left(\vec{r} - \vec{r}' \right)} \sin\alpha d\alpha d\beta = \frac{e^{-ik|\vec{r} - \vec{r}'|}}{4\pi|\vec{r} - \vec{r}'|}. \qquad (1.20)$$

The current support in relations (1.17)–(1.19) can, in particular, be a surface (closed or nonclosed). If we deal with the problem of wave diffraction at a compact scatterer occupying a spatial domain D with boundary S, then, according to (1.19), the field $\vec{E}^1, \vec{H}^1$ scattered at an obstacle can be represented everywhere outside S as

$$\vec{E}^1(\vec{r}) = \int_S \left\{ \frac{\zeta}{ik} \left[\nabla \times \left(\nabla \times \left(\vec{n}' \times \vec{H}^1 \right) G_0 \right) \right] - \left(\nabla \times \left(\vec{E}^1 \times \vec{n}' \right) G_0 \right) \right\} ds',$$

$$\vec{H}^1(\vec{r}) = \int_S \left\{ \frac{1}{ik\zeta} \left[\nabla \times \left(\nabla \times \left(\vec{E}^1 \times \vec{n}' \right) G_0 \right) \right] + \left(\nabla \times \left(\vec{n}' \times \vec{H}^1 \right) G_0 \right) \right\} ds', \quad (1.21)$$

where $\vec{n}'$ is the vector of the outer normal on the surface S at the point of integration.

Let A be the set of singularities of the diffraction field $\vec{E}^1, \vec{H}^1$, and let Σ be an arbitrary sufficiently smooth surface surrounding $\bar{A}$. Then it follows from the equivalence principle [31] that $\vec{E}^1, \vec{H}^1$ can be considered as the field of currents $\vec{J}^e = \left(\vec{n} \times \vec{H}^1\right)\big|_\Sigma$ and $\vec{J}^m = \left(\vec{E}^1 \times \vec{n}\right)\big|_\Sigma$ localized on Σ. Therefore, repeating the argument prior to formulas (1.19), we obtain the representations

$$\vec{E}^1\left(\vec{r}\right) = \int_\Sigma \left\{ \frac{\zeta}{ik}\left[\nabla \times \left(\nabla \times \left(\vec{n}' \times \vec{H}^1\right)G_0\right)\right] - \left(\nabla \times \left(\vec{E}^1 \times \vec{n}'\right)G_0\right)\right\}d\sigma',$$

$$\vec{H}^1\left(\vec{r}\right) = \int_\Sigma \left\{ \frac{1}{ik\zeta}\left[\nabla \times \left(\nabla \times \left(\vec{E}^1 \times \vec{n}'\right)G_0\right)\right] + \left(\nabla \times \left(\vec{n}' \times \vec{H}^1\right)G_0\right)\right\}d\sigma'. \quad (1.22)$$

If the surface S is analytic, then the surface Σ can strictly lie inside S. Let Ω be the domain between S and Σ. Repeating the argument prior to formulas (1.7) but with significantly more cumbersome calculations [5], we can show that

$$\int_S \left\{ \frac{\zeta}{ik}\left[\nabla \times \left(\nabla \times \left(\vec{n}' \times \vec{H}^1\right)G_0\left(\vec{r};\vec{r}'\right)\right)\right] - \left(\nabla \times \left(\vec{E}^1 \times \vec{n}'\right)G_0\left(\vec{r};\vec{r}'\right)\right)\right\}ds' = 0,$$

$$\int_S \left\{ \frac{1}{ik\zeta}\left[\nabla \times \left(\nabla \times \left(\vec{E}^1 \times \vec{n}'\right)G_0\left(\vec{r};\vec{r}'\right)\right)\right] + \left(\nabla \times \left(\vec{n}' \times \vec{H}^1\right)G_0\left(\vec{r};\vec{r}'\right)\right)\right\}ds' = 0,$$

$$(1.23)$$

for $M\left(\vec{r}\right) \in \Omega$.

As in the scalar case (see above), since the integrals on the left-hand side in (1.23) are analytic everywhere inside S, these relations can be rewritten as

$$\int_S \left\{ \frac{\zeta}{ik}\left[\nabla \times \left(\nabla \times \left(\vec{n}' \times \vec{H}^1\right)G_0\left(\vec{r};\vec{r}'\right)\right)\right]\right.$$

$$\left. - \left(\nabla \times \left(\vec{E}^1 \times \vec{n}'\right)G_0\left(\vec{r};\vec{r}'\right)\right)\right\}ds' = 0, \quad M\left(\vec{r}\right) \in D,$$

$$\int_S \left\{ \frac{1}{ik\zeta}\left[\nabla \times \left(\nabla \times \left(\vec{E}^1 \times \vec{n}'\right)G_0\left(\vec{r};\vec{r}'\right)\right)\right]\right.$$

$$\left. + \left(\nabla \times \left(\vec{n}' \times \vec{H}^1\right)G_0\left(\vec{r};\vec{r}'\right)\right)\right\}ds' = 0, \quad M(\vec{r}) \in D. \quad (1.24)$$

As previous, taking into account the relations

$$\int_S \left\{ \frac{\zeta}{ik} \left[\nabla \times \left(\nabla \times \left(\vec{n}' \times \vec{H}^0 \right) G_0 \left(\vec{r}; \vec{r}' \right) \right) \right] \right.$$

$$\left. - \left(\nabla \times \left(\vec{E}^0 \times \vec{n}' \right) G_0 \left(\vec{r}; \vec{r}' \right) \right) \right\} ds' = \begin{cases} 0, & M(\vec{r}) \in R^3 \backslash \bar{D}, \\ -\vec{E}^0(\vec{r}), & M(\vec{r}) \in D, \end{cases}$$

$$\int_S \left\{ \frac{1}{ik\zeta} \left[\nabla \times \left(\nabla \times \left(\vec{E}^0 \times \vec{n}' \right) G_0 \left(\vec{r}; \vec{r}' \right) \right) \right] \right.$$

$$\left. + \left(\nabla \times \left(\vec{n}' \times \vec{H}^0 \right) G_0 \left(\vec{r}; \vec{r}' \right) \right) \right\} ds' = \begin{cases} 0, & M(\vec{r}) \in R^3 \backslash \bar{D}, \\ -\vec{H}^0(\vec{r}), & M(\vec{r}) \in D, \end{cases}$$

where $\vec{E}^0, \vec{H}^0$ is the initial field, formulas (1.24) can be written as [5]:

$$\int_S \left\{ \frac{\zeta}{ik} \left[\nabla \times \left(\nabla \times \left(\vec{n}' \times \vec{H} \right) G_0 \left(\vec{r}; \vec{r}' \right) \right) \right] \right.$$

$$\left. - \left(\nabla \times \left(\vec{E} \times \vec{n}' \right) G_0 \left(\vec{r}; \vec{r}' \right) \right) \right\} ds' = \begin{cases} \vec{E}^1(\vec{r}), & M(\vec{r}) \in R^3 \backslash \bar{D}, \\ -\vec{E}^0(\vec{r}), & M(\vec{r}) \in D, \end{cases}$$

$$\int_S \left\{ \frac{1}{ik\zeta} \left[\nabla \times \left(\nabla \times \left(\vec{E}^0 \times \vec{n}' \right) G_0 \left(\vec{r}; \vec{r}' \right) \right) \right] \right.$$

$$\left. + \left(\nabla \times \left(\vec{n}' \times \vec{H}^0 \right) G_0 \left(\vec{r}; \vec{r}' \right) \right) \right\} ds' = \begin{cases} \vec{H}^1(\vec{r}), & M(\vec{r}) \in R^3 \backslash \bar{D}, \\ -\vec{H}^0(\vec{r}), & M(\vec{r}) \in D, \end{cases}$$

$$(1.25)$$

where $\vec{E} = \vec{E}^0 + \vec{E}^1$, $\vec{H} = \vec{H}^0 + \vec{H}^1$ is now the complete (incident plus scattered) field. It may turn out that formulas (1.25) are more preferable than (1.24), because they permit taking the boundary conditions on S into account.

By analogy with the scalar (acoustic) case, representations (1.22) and (1.25) are also called wave potentials.

1.1.2 Representation by a Series in Wave Harmonics and the Atkinson-Wilcox Expansion

The above representation can easily be obtained by the operator method [34], which is based on a very simple idea. Let us consider the linear second-order partial differential equation

$$\sum_{j=1}^{n} a_j \frac{\partial^2 u}{\partial x_j^2} + \sum_{j=1}^{n} b_j \frac{\partial u}{\partial x_j} + cu = f(x_1,\ldots,x_n). \tag{1.26}$$

In Equation (1.26), we distinguish one variable, for example, x_1, and assume that the differentiation operators with respect to other variables are parameters. As a result, we obtain a linear ordinary differential equation with operator coefficients. Its formal solution can easily be obtained in many cases. Now the main difficulty in the problem is to justify the obtained solution. Several examples of this method application to equations of different types (hyperbolic, elliptic, and parabolic) (e.g., see [34]).

First, we consider the two-dimensional homogeneous Helmholtz equation in polar coordinates

$$\frac{\partial^2 U^1}{\partial r^2} + \frac{1}{r} \frac{\partial U^1}{\partial r} + \frac{1}{r^2} \frac{\partial^2 U^1}{\partial \varphi^2} + k^2 U^1 = 0.$$

We rewrite it as

$$\left[\frac{\partial^2}{\partial r^2} + \frac{1}{r} \frac{\partial}{\partial r} + k^2 - \frac{(-i \partial/\partial \varphi)^2}{r^2} \right] U^1(r, \varphi) = 0.$$

Assume that the operator $(-i(\partial/\partial \varphi)) \equiv \nu$ is a parameter, and consider the ordinary differential equation

$$\frac{d^2 H^\nu}{dr^2} + \frac{1}{r} \frac{dH^\nu}{dr} + \left(k^2 - \frac{\nu^2}{r^2} \right) H^\nu = 0,$$

which is the Bessel equation. Its general solution has the form

$$H^\nu = e^{-i\nu(\pi/2)} H_\nu^{(2)}(kr) C_1(\varphi) + e^{i\nu(\pi/2)} H_\nu^{(1)}(kr) C_2(\varphi),$$

where $H_\nu^{(1)}(kr)$ and $H_\nu^{(2)}(kr)$ are two linearly independent solutions of the Bessel equation (the Hankel functions of the first and second type, respectively). If the harmonic dependence on time is chosen in the form $e^{i\omega t}$, then only the function $H_\nu^{(2)}(kr)$ satisfies the Sommerfeld condition at infinity (the radiation condition). Indeed, as $kr \to \infty$, we have

$$e^{-i\nu(\pi/2)} H_\nu^{(2)}(kr) = \sqrt{\frac{2}{\pi kr}} e^{-ikr + i(\pi/4)} \left(1 + O\left(\frac{1}{kr} \right) \right),$$

and hence

$$U^1(r, \varphi) e^{i\omega t} = \sqrt{\frac{2}{\pi kr}} e^{i(\pi/4)} e^{i(\omega t - kr)} C_1(\varphi). \tag{1.27}$$

This is a cylindrical wave propagating with the phase velocity

$$v_p \equiv \frac{dr}{dt} = \frac{\omega}{k} = c.$$

The term with $H_\nu^{(1)}(kr)$ corresponds to the wave traveling in the negative direction ($v_p = -\frac{\omega}{k}$). The function $C_1(\varphi)$ has the meaning of the wave field pattern (see below), and hence we introduce the notation $C_1(\varphi) \equiv f(\varphi)$.

If we represent the operator function $e^{-i\nu(\pi/2)} H_\nu^{(2)}(kr)$ as

$$e^{-i\nu(\pi/2)} H_\nu^{(2)}(kr) \sim \sqrt{\frac{2}{\pi kr}} e^{-ikr + i(\pi/4)} \sum_{n=0}^{\infty} \left(\frac{i}{2kr}\right)^n \frac{1}{n!} \prod_{m=1}^{n} \left[\left(m - \frac{1}{2}\right)^2 - \nu^2\right],$$

which follows from the corresponding asymptotic expansion of the function $e^{-i\nu(\pi/2)} H_\nu^{(2)}(kr)$ [35], then we obtain the asymptotic Atkinson-Wilcox series [5,36,37] for $U^1(r, \varphi)$, which is widely used, in particular, in the asymptotic theory of diffraction [38],

$$U^1(r, \varphi) \sim \sqrt{\frac{2}{\pi kr}} e^{-ikr + i(\pi/4)} \sum_{n=0}^{\infty} \frac{a_n(\varphi)}{(kr)^n}. \tag{1.28}$$

where

$$a_n(\varphi) = \frac{i}{2n} \left[\left(n - \frac{1}{2}\right)^2 + \left(\frac{d}{d\varphi}\right)^2\right] a_{n-1}(\varphi), \quad n = 1, 2, 3, \ldots$$

Comparing (1.28) with (1.27), we see that the function $a_0(\varphi)$ has the meaning of the wave field pattern, i.e.,

$$a_0(\varphi) \equiv f(\varphi).$$

Since $e^{-i\nu(\pi/2)} H_\nu^{(2)}(kr)$ is an entire function in the variable ν [35], we can use the representation

$$e^{-i\nu\frac{\pi}{2}} H_\nu^{(2)}(kr) = \frac{1}{2\pi i} \oint_C \frac{e^{-i\zeta\frac{\pi}{2}} H_\zeta^{(2)}(kr)}{\zeta - \nu} d\zeta,$$

where C is a circle whose radius tends to infinity.

We consider the action of the operator $\dfrac{1}{\zeta + i(d/d\varphi)} \equiv \dfrac{1}{\zeta - \nu}$ on the function $f(\varphi)$, which we represent by the Fourier series

$$\frac{1}{\zeta + i\frac{d}{d\varphi}} f(\varphi) = \frac{1}{\zeta} \sum_{n=0}^{\infty} \left(\frac{-i}{\zeta}\right)^n f^{(n)}(\varphi) = \frac{1}{\zeta} \sum_{m=-\infty}^{\infty} a_m e^{im\varphi} \sum_{n=0}^{\infty} \left(-\frac{i}{\zeta} im\right)^n$$

$$= \sum_{m=-\infty}^{\infty} \frac{a_m e^{im\varphi}}{\zeta - m},$$

where a_m are the Fourier coefficients of the function $f(\varphi)$. Thus, we have

$$U^1(r, \varphi) = e^{-i\nu\frac{\pi}{2}} H_\nu^{(2)}(kr) f(\varphi) = \sum_{m=-\infty}^{\infty} a_m e^{im\varphi} \frac{1}{2\pi i} \oint_C \frac{e^{-i\zeta\frac{\pi}{2}} H_\zeta^{(2)}(kr)}{\zeta - m} d\zeta$$

$$= \sum_{m=-\infty}^{\infty} a_m e^{-im\frac{\pi}{2}} H_m^{(2)}(kr) e^{im\varphi}. \tag{1.29}$$

The obtained formula is the Rayleigh series for the wave field in the two-dimensional case.

Now we consider the three-dimensional Helmholtz equation in the form

$$\frac{1}{kr} \frac{\partial^2 (krU^1)}{\partial (kr)^2} + \frac{1}{(kr)^2} DU^1 + U^1 = 0, \tag{1.30}$$

where

$$D \equiv \frac{1}{\sin\theta} \frac{\partial}{\partial\theta} \left(\sin\theta \frac{\partial}{\partial\theta} \right) + \frac{1}{\sin^2\theta} \frac{\partial^2}{\partial\varphi^2}$$

is the Beltrami operator [6].

We introduce the operator B:

$$B(B+1) = -D$$

and determine the operator function $H^B(kr)$ from the equation

$$\frac{1}{kr} \frac{d^2(krH^B)}{d(kr)^2} + \left[1 - \frac{B(B+1)}{(kr)^2} \right] H^B = 0. \tag{1.31}$$

The solution of the Bessel equation (1.31) satisfying the radiation condition has the form

$$H^B = e^{-iB\frac{\pi}{2}} h_B^{(2)}(kr) = \frac{i}{kr} \sum_{p=0}^{\infty} \frac{(-1)^p e^{-ikr}}{p!(2ikr)^p} \prod_{n=1}^{p} [n(n-1) - B(B+1)], \tag{1.32}$$

where $h_B^{(2)}(kr) = \sqrt{2/\pi kr} H_{B+1/2}^{(2)}(kr)$ is the spherical Hankel function of the second type [4].

As previously, to solve Equation (1.30), it is necessary to act by an operator function on a function $f(\theta, \varphi)$, which we write as

$$f(\theta, \varphi) = \sum_{n=0}^{\infty} \sum_{m=-n}^{n} a_{nm} P_n^m(\cos\theta) e^{im\varphi},$$

where $P_n^m(\cos\theta)$ are associated Legendre functions [4,39].

We use the relations [39]

$$DP_n^m(\cos\theta)e^{im\varphi} = -n(n+1)P_n^m(\cos\theta)e^{im\varphi}$$

to obtain

$$U^1(r,\theta,\varphi) \equiv H^B f(\theta,\varphi) = i\frac{e^{-ikr}}{kr}\sum_{p=0}^{\infty}\frac{(-1)^p}{(2ikr)^p}$$

$$\times\sum_{n=0}^{\infty}\sum_{m=-n}^{n}a_{nm}\frac{1}{p!}\prod_{q=1}^{p}[q(q-1)-n(n+1)]P_n^m(\cos\theta)e^{im\varphi}$$

It is well known [35] that

$$h_n^{(2)}(kr) = \frac{ie^{-ikr+in(\pi/2)}}{kr}\sum_{p=0}^{n}\frac{(n+1/2,p)}{(2ikr)^p},$$

where

$$\left(n+\frac{1}{2},p\right) \equiv \frac{1}{p!}\prod_{q=1}^{p}[q(q-1)-n(n+1)]$$

is the Hankel symbol.

Thus, we finally obtain the relation given in the "Introduction"

$$U^1(r,\theta,\varphi) = \sum_{n=0}^{\infty}\sum_{m=-n}^{n}a_{nm}(-i)^{n+1}h_n^{(2)}(kr)P_n^m(\cos\theta)e^{im\varphi}. \tag{1.33}$$

The expansion in the wave harmonics (Rayleigh series) in the three-dimensional case is obtained.

In the domain of convergence of the series (1.33), the terms in this series can be regrouped so as to obtain the Atkinson-Wilcox expansion [36]

$$U^1(r,\theta,\varphi) = \frac{e^{-ikr}}{kr}\sum_{p=0}^{\infty}\frac{a_p(\theta,\varphi)}{(kr)^p}, \tag{1.34}$$

where

$$a_0(\theta,\varphi) = f(\theta,\varphi), \quad a_p(\theta,\varphi) = \frac{i}{2p}[p(p-1)+D]a_{p-1}(\theta,\varphi), \quad p=1,2,\ldots$$

This approach can be completely transferred to the vector case.

1.1.3 Integral and Series of Plane Waves

Now we use the following integral representation for the operator function $e^{-i\nu(\pi/2)}H_\nu^{(2)}(kr)$,

$$e^{-i\nu\frac{\pi}{2}}H_\nu^{(2)}(kr) = \frac{1}{\pi}\int_{-(\pi/2)-i\infty}^{\pi/2+i\infty} \exp(-ikr\cos\psi + i\nu\psi)\,d\psi,$$

and obtain

$$U^1(r,\varphi) = \frac{1}{\pi}\int_{-(\pi/2)-i\infty}^{\pi/2+i\infty} \exp(-ikr\cos\psi)e^{\psi(d/d\varphi)}f(\varphi)\,d\psi$$

for the wave field. Here $e^{\psi(d/d\varphi)}$ is the shift operator, i.e., $e^{\psi(d/d\varphi)}f(\varphi) = f(\varphi+\psi)$, and hence we have

$$U^1(r,\varphi) = \frac{1}{\pi}\int_{-(\pi/2)-i\infty}^{\pi/2+i\infty} \exp(-ikr\cos\psi)f(\varphi+\psi)\,d\psi. \tag{1.35}$$

The obtained relation is the generalized Sommerfeld representation [34]. The standard representation has the form [33,40]

$$U^1(r,\varphi) = \frac{1}{\pi}\int_{-i\infty}^{\pi+i\infty} f(\pm\psi)\exp[-ikr\cos(\varphi\mp\psi)]\,d\psi, \tag{1.36}$$

where the upper sign is taken for $y \equiv r\sin\varphi > a$, and the lower, for $r\sin\varphi < -b$, a is the maximal ordinate of the scatterer cross-section contour, and $-b$ is the minimal ordinate.

Below we show that representation (1.35) is more convenient in several cases where the diffraction problems are solved, because for a sufficiently wide class of scatterer geometries, this representation converges till the scatterer boundary, while representation (1.36) is meaningful only in the half-planes above (for the upper signs) and below (for the lower signs) the scatterer. In what follows, we specify the boundaries of the domains of convergence of representations (1.35) and (1.36).

Similarly, in the three-dimensional case, the wave field representation by the integral of plane waves (Sommerfeld–Weyl integral) has the form [10,41]

$$U^1(r,\theta,\varphi) = \frac{1}{2\pi i}\int_0^{2\pi} d\beta \int_0^{\pi/2+i\infty} f(\alpha,\beta)\exp(-ikr\cos\gamma)\sin\alpha\,d\alpha, \tag{1.37}$$

where

$$f(\alpha,\beta) = \frac{k}{4\pi}\int_S \left\{ U(\vec{r}')\frac{\partial}{\partial n'} - \frac{\partial U(\vec{r}')}{\partial n'} \right\} \exp(ikr'\cos\gamma')\,ds',$$

and

$$\cos\gamma = \sin\theta \sin\alpha \cos(\varphi - \beta) + \cos\theta \cos\alpha \equiv \vec{p} \cdot \vec{r},$$
$$\cos\gamma' = \sin\theta' \sin\alpha \cos(\varphi' - \beta) + \cos\theta' \cos\alpha \equiv \vec{p} \cdot \vec{r'},$$
$$\vec{p} = \{\sin\alpha\cos\beta, \sin\alpha\sin\beta, \cos\alpha\}.$$

The integral (1.37) converges for $z \equiv r\sin\theta > z_0$, where the value z_0 will be determined later (in any case, it does not exceed the maximal applicate of the domain D). This is inconvenient when we solve the boundary-value problems, and therefore, as in the two-dimensional case, it is expedient to obtain the generalized representation (1.37), which converges in the domain outside the scatterer.

To obtain the generalized representation, it is more convenient to solve the boundary-value problems as follows [42]. We rotate the coordinate system so that the axis Oz is directed towards the observation point. For this, it is necessary to rotate the system first by the angle $\frac{\pi}{2} + \varphi$ about the axis Oz and then by the angle θ about the new axis Ox. We apply the matrix of rotation

$$B = \begin{pmatrix} -\sin\varphi & \cos\varphi & 0 \\ -\cos\varphi\cos\theta & -\sin\varphi\cos\theta & \sin\theta \\ \cos\varphi\sin\theta & \sin\varphi\sin\theta & \cos\theta \end{pmatrix}$$

to the vectors $\vec{r}$ and $\vec{r'}$. As a result, we obtain

$$B\vec{r} = \begin{pmatrix} 0 \\ 0 \\ r \end{pmatrix}, \quad B\vec{r'} = r' \begin{pmatrix} \sin\theta'\sin(\varphi - \varphi') \\ -\cos\theta\sin\theta'\cos(\varphi - \varphi') + \cos\theta'\sin\theta \\ \sin\theta\sin\theta'\cos(\varphi - \varphi') + \cos\theta'\cos\theta \end{pmatrix}.$$

Now the wave field representation by the integral of plane waves has the form

$$U^1(r, \theta, \varphi) = \frac{1}{2\pi i} \int_0^{2\pi} \int_0^{\pi/2 + i\infty} \exp(-ikr\cos\alpha)\hat{f}(\alpha, \beta; \theta, \varphi)\sin\alpha \, d\alpha \, d\beta.$$

$$(1.38)$$

Below we show that representation (1.38) holds everywhere outside the convex envelope of singularities of the wave field continuation, and hence for a sufficiently wide class of geometries, i.e., everywhere to the scatterer boundary. This allows us to solve the boundary-value problems by using this representation. In representation (1.38),

$$\hat{f}(\alpha, \beta; \theta, \varphi) = \frac{k}{4\pi} \int_S \left\{ U(\vec{r''})\frac{\partial}{\partial n'} - \frac{\partial U(\vec{r''})}{\partial n'} \right\} \exp\left(ik\vec{p} \cdot B\vec{r'} \right) ds'. \quad (1.39)$$

is a generalized wave field pattern. We note that such generalized representations also hold in the vector case (see [111]). At the end of this section, we consider the wave field representation in the form of a series of plane waves [9, Chapter 2]. Such representations are used to solve the problems of wave diffraction at periodic structures (e.g., see [43–46]).

In the two-dimensional case, such a representation has the form [45], [9, Chapter 2]

$$U^1(x, y) = \frac{2}{b} \sum_{n=-\infty}^{\infty} f_0(w_n) \frac{\exp(-iw_n x)\exp(-iv_n y)}{v_n}, \qquad (1.40)$$

where $f_0(w_n)$ is the central period pattern [45]

$$f_0(w) = \frac{1}{4i} \int_{-(b/2)}^{b/2} \left(U(\vec{r}) \frac{\partial}{\partial n} - \frac{\partial U(\vec{r})}{\partial n} \right) \exp[iwx + ivh(x)] \sqrt{1 + h'^2(x)} dx. \qquad (1.41)$$

or $f_0(w) = f\left(\arccos \frac{w}{k}\right)$ in the case of diffraction at a discrete periodic grating, and $f(\psi)$ is determined by the relation [9, Chapter 2]

$$f(\psi) = \frac{1}{4i} \int_S \left\{ U(\vec{r}) \frac{\partial}{\partial n} - \frac{\partial U(\vec{r})}{\partial n} \right\} \exp(ikr\cos(\psi - \varphi)) ds.$$

The coefficients at the mth spectral orders are related to $f_0(w_m)$ as

$$R_m = \frac{2}{b} \frac{f_0(w_m)}{v_m}. \qquad (1.42)$$

The quantities w_m and v_m are discussed after formula (1.12).

In the three-dimensional case, the corresponding representation is quite similar (e.g., see [46]).

1.2 ANALYTIC PROPERTIES OF THE WAVE FIELD PATTERN AND THE DOMAINS OF EXISTENCE OF ANALYTIC REPRESENTATIONS

1.2.1 Analytic Properties of the Wave Field Pattern

Now we return to representation (1.2). In the so-called far-field region, i.e., for $kr \gg 1$, $r \gg d$, where d is the maximal diameter of Σ, the function $G_0(\vec{r}; \vec{r}')$ can be written as

$$G_0\left(\vec{r};\vec{r'}\right) = \frac{e^{-ikr}}{4\pi r}\exp\{ikr'[\sin\theta\sin\theta'\cos(\varphi-\varphi') + \cos\theta\cos\theta']\} + O\left(\frac{1}{r^2}\right)$$

in the three-dimensional case and as

$$G_0\left(\vec{r};\vec{r'}\right) = \sqrt{\frac{2}{\pi kr}}e^{-ikr+i\pi/4}\exp\{ikr'\cos(\varphi-\varphi')\} + O\left(\frac{1}{r^{3/2}}\right)$$

in the two-dimensional case. Then formula (1.2) implies

$$U^1\left(\vec{r}\right) = \frac{e^{-ikr}}{r}f(\theta,\varphi) + O\left(\frac{1}{r^2}\right),$$

$$U^1\left(\vec{r}\right) = \sqrt{\frac{2}{\pi kr}}e^{-ikr+i\pi/4}f(\varphi) + O\left(\frac{1}{r^{3/2}}\right)$$

in the three- and two-dimensional cases, respectively.

In these relations, we have

$$f(\alpha,\beta) = \frac{1}{4\pi}\int_\Sigma\left\{U^1\left(\vec{r}\right)\frac{\partial}{\partial n} - \frac{\partial U^1\left(\vec{r}\right)}{\partial n}\right\}e^{ikr[\sin\alpha\sin\theta\cos(\beta-\varphi) + \cos\alpha\cos\theta]}d\sigma,$$

$$f(\psi) = \frac{1}{4i}\int_\Sigma\left\{U^1\left(\vec{r}\right)\frac{\partial}{\partial n} - \frac{\partial U^1\left(\vec{r}\right)}{\partial n}\right\}e^{ikr\cos(\psi-\varphi)}d\sigma.$$

$$(1.43)$$

We note that, by (1.11) the function $U^1\left(\vec{r}\right)$ in integrals (1.43) can be replaced by $U(\vec{r})$.

The functions $f(\alpha,\beta)$ and $f(\psi)$ are called wave field patterns, and they show how the diffraction field depends on the angular coordinates in the far-field region.

Let us study the analytic properties of these functions. First, we consider the two-dimensional case. We set $\psi = \alpha + i\beta$. It is easy to see that

$$2\cos(\psi-\varphi) = R\left[\cos(\varphi-\alpha) + is_\beta\sin(\varphi-\alpha)\right] + \frac{1}{R}\left[\cos(\varphi-\alpha) - is_\beta\sin(\varphi-\alpha)\right],$$

$$(1.44)$$

where $R = e^{|\beta|}$, $s_\beta = \text{sign}\beta$.

Formulas (1.43) and (1.44) imply the asymptotic relation as $R \to \infty$

$$f(\psi) = f^E(w_\pm)\left(1 + O\left(\frac{1}{R}\right)\right),$$

$$(1.45)$$

where

$$f^E(w_\pm) = \frac{1}{4i}\int_\Sigma \left\{ U^1(\vec{r})\frac{\partial}{\partial n} - \frac{\partial U^1(\vec{r})}{\partial n}\right\} \exp\left(\frac{ikr}{2}Re^{is_\beta(\varphi-\alpha)}\right)d\sigma, \quad (1.46)$$

is an entire function of a finite degree of the variables $w_\pm = Re^{\pm i\alpha}$, the upper sign in (1.46) corresponds to $s_\beta = -1$, and the lower sign corresponds to $s_\beta = +1$.

We assume that the equation of the contour Σ in polar coordinates has the form $r = \rho_\Sigma(\varphi)$. Because

$$\frac{\partial}{\partial n} = \left(\rho_\Sigma^2 + \rho_\Sigma'^2\right)^{-1/2}\left(\rho_\Sigma(\varphi)\frac{\partial}{\partial r} - \frac{\rho_\Sigma'(\varphi)}{\rho_\Sigma(\varphi)}\frac{\partial}{\partial \varphi}\right), \quad d\sigma = \left(\rho_\Sigma^2 + \rho_\Sigma'^2\right)^{1/2}d\varphi,$$

we can rewrite (1.46) as

$$f^E(w_\pm) = \frac{1}{4i}\int_0^{2\pi}\left\{\left[\rho_\Sigma(\varphi)\frac{\partial U^1}{\partial r} - \frac{\rho_\Sigma'(\varphi)}{\rho_\Sigma(\varphi)}\frac{\partial U^1}{\partial \varphi}\right] - s_\beta\frac{kw_\pm}{2}U^1\left[\rho_\Sigma(\varphi)e^{is_\beta\varphi}\right]'\right\}$$
$$\times \exp\left(\frac{ik\rho_\Sigma(\varphi)}{2}w_\pm e^{is_\beta\varphi}\right)d\varphi.$$

$$(1.47)$$

Now let us estimate the integral (1.47), where we set $w_\pm = w_+ \equiv w$ and $f^E(w_\pm) = f^E(w)$, by using the Cauchy-Bunyakovsky inequality

$$\left|f^E(w)\right|^2 \le \|I\|^2 \int_0^{2\pi}\exp[Rk\rho_\Sigma(\varphi)\sin(\alpha-\varphi)]d\varphi, \quad (1.48)$$

where

$$\|I\|^2 = \frac{1}{16}\int_0^{2\pi}\left|\left[\rho_\Sigma(\varphi)\frac{\partial U^1}{\partial r} - \frac{\rho_\Sigma'(\varphi)}{\rho_\Sigma(\varphi)}\frac{\partial U^1}{\partial \varphi}\right] + \frac{kw}{2}U^1\left[\rho_\Sigma(\varphi)e^{-is\varphi}\right]'\right|^2 d\varphi.$$

$$(1.49)$$

Under the assumption that $\|I\|^2$ is proportional to $|w|$, we use inequality (1.48) to obtain

$$\max_{|w|=R}|f^E(w)| \le \text{const}_1 R\exp\left(R\frac{kr_0}{2}\right) < \exp\left[R\left(\frac{kr_0}{2} + \varepsilon_1\right)\right],$$
$$\max_{w=\pm R}|f^E(w)| \le \text{const}_2 R\exp\left(R\frac{ka}{2}\right) < \exp\left[R\left(\frac{ka}{2} + \varepsilon_2\right)\right],$$

$$(1.50)$$

where $r_0 = \max_{\varphi}\rho_{\Sigma}(\varphi)$ is the radius of the circle circumscribed about Σ, $a = \max_{\varphi}[\rho_{\Sigma}(\varphi)\sin\varphi]$ is the maximal ordinate of the contour Σ, and $\varepsilon_1, \varepsilon_2$ are arbitrarily small positive numbers. The second inequalities in (1.50) hold as $R \to \infty$.

Thus, it follows from estimates (1.50) that (because $\varepsilon_1, \varepsilon_2$ are arbitrarily small)

$$\varlimsup_{R\to\infty} \frac{\ln \max_{|w|=R}|f^E(w)|}{R} = \sigma, \quad \sigma \le \frac{kr_0}{2}, \tag{1.51}$$

$$\varlimsup_{R\to\infty} \frac{\ln \max_{w=\pm R}|f^E(w)|}{R} = \sigma_S, \quad \sigma_S \le \frac{ka}{2}. \tag{1.52}$$

It also follows from (1.51) that $f^E(w)$ is a first-order entire function of a finite degree [5,9,47] σ, which does not exceed $kr_0/2$. Relation (1.52) is the definition of the degree of the function $f^E(w)$ on the Sommerfeld contour (i.e., for $\alpha = 0$, $s_\beta = -1$ and $\alpha = \pi$, $s_\beta = 1$).

We note that, as follows from definitions (1.51) and (1.52), the value of σ is invariant under rotations of the coordinate system but depends on the origin location [9, Chapter 2]. Conversely, the value of σ_S varies with rotations of the coordinate system but is independent of the origin displacement along the axis Ox, and in the case of displacements along the axis Oy, the variation in the value of $\frac{2}{k}\sigma_S$ is exactly equal to the variation in the coordinate y. In particular, if the coordinate axes are turned by an angle γ in the positive direction, then we have [9, Chapter 2]

$$\sigma_S(\gamma) = \varlimsup_{R\to\infty} \frac{\ln \max_{w=\pm R}|f^E(we^{-is_\beta\gamma})|}{R}. \tag{1.52a}$$

Thus, we have shown that if the function $f(\psi)$ is a wave field pattern and can therefore be represented by the integral (1.43), then it can analytically be continued to the entire complex plane $\psi = \alpha + i\beta$ and asymptotically coincides (see (1.45)) with an entire function of a finite degree of complex variables $w_{\pm} = \exp(|\beta| \pm i\alpha)$ near the infinitely remote point (as $|\beta| \to \infty$).

Because of the asymptotic relation (1.45), the degrees σ and σ_S of the function $f^E(w)$ are also called the corresponding degrees of the pattern. The following remark is appropriate here. It is well known [5,47] that the directivity pattern of a plane-aperture antenna is an entire function of a finite degree. In contrast, the scattering pattern of a compact body bounded by a closed surface (by a closed curve in the two-dimensional case) is generally not an entire function of exponential type but only coincides (in the sense of asymptotic relation (1.45)) with an entire function of a finite degree in a neighborhood of the infinitely remote point on the complex plane of its argument. The only exception is formed by plane scatterers, i.e., by the scatterers whose "boundary" S is a part of the plane (a segment of a line in the two-dimensional case). The pattern of such scatterers are also entire functions of exponential type.

In the three-dimensional case, the asymptotic relation [10,41]

$$f(\theta, \varphi) = f^E(w_\mp, \varphi)\left(1 + O\left(e^{-|\theta_2|}\right)\right), \quad w_\mp = \exp(|\theta_2| \mp i\theta_1),$$

which is similar to (1.45), holds on the complex plane $\theta = \theta_1 + i\theta_2$ for the wave field pattern $f(\theta, \varphi)$. As in the two-dimensional case, $f^E(w_\mp, \varphi)$ is an entire function in the variables $w_\mp$ of a finite degree σ less than or equal to $\frac{kr_0}{2}$, where r_0 is the radius of the sphere circumscribed around Σ. As in the two-dimensional case, σ is called the degree of the pattern $f(\theta, \varphi)$.

The function $f^E(w_\pm)$ can be represented everywhere by the converging series

$$f^E(w_\pm) = \sum_{n=0}^{\infty} c_n w_\pm^n \tag{1.53}$$

whose coefficients satisfy the limit relation

$$\overline{\lim_{n \to \infty}} \sqrt[n]{n!|c_n|} = \sigma \tag{1.54}$$

due to the following theorem [5,47]:

Theorem 1 *$f(z)$ is an entire function of a finite degree σ if and only if the coefficients c_n of its expansion in a power series satisfy relation (1.54).*

To determine σ and σ_S, we estimate the integral (1.47), where $\rho_\Sigma(\varphi)$ is the equation of the curve Σ surrounding the set of singularities $\bar{A}$ of the analytic continuation of the wave field. We can assume that Σ surrounds the set $\bar{A}$ extremely closely. Then r_0 in the inequality $\sigma \leq \frac{kr_0}{2}$ is the distance from the origin to the most remote point of the set A and a in the inequality $\sigma_S \leq \frac{ka}{2}$ is the ordinate of the point in this set that is most remote from the abscissa axis. Thus, to determine σ and σ_S, it is necessary to know the geometry of the set A of the wave field singularities [48].

1.2.2 Localization of Singularities of the Wave Field Analytic Continuation

We consider representation (1.7) of the wave field scattered at an obstacle bounded by a surface (curve) S. If the scatterer boundary S is analytic, i.e., if its equation is a real analytic function of coordinates, then the field $U^1(\vec{r})$ can analytically be continued by an analytic deformation of S. Let us consider the technique of such a deformation used to solve the two-dimensional problem of diffraction at a compact scatterer. We assume that the equation of the scatterer boundary S can be written as $r = \rho(\varphi)$ (for simplicity, we assume that this boundary is star-shaped). We introduce the complex variable

$$\zeta = \rho(\varphi)e^{i\varphi}. \tag{1.55}$$

If φ in (1.55) is real, then on the complex plane $z = re^{i\varphi}$, the variable ζ circumscribes a contour C that geometrically coincides with S. But if we set $\varphi = \varphi_1 + i\varphi_2$, then the contour C is deformed, in particular, it contracts for positive values of φ_2.

Such a deformation is possible until the mapping of (1.55) ceases to be - one-to-one. Moreover, since the boundary values of the functions $U^1(\vec{r})|_S$ and $\frac{\partial U^1(\vec{r})}{\partial n}|_S$ are related to the values of the functions $U^0(\vec{r})|_S$ and $\frac{\partial U^0(\vec{r})}{\partial n}|_S$ through the boundary conditions, such a deformation is also bounded by the singular points of the function $U^0(\vec{r})|_S$ when it is continued to the region of complex φ.

Obviously, the mapping (1.55) ceases to be one-to-one at the points, where

$$\zeta'(\varphi) \equiv [\rho'(\varphi) + i\rho(\varphi)]e^{i\varphi} = 0. \tag{1.56}$$

It should be noted that Equation (1.56) is equivalent to the set of two equations

$$\begin{bmatrix} [\rho'(\varphi) + i\rho(\varphi)] = 0, \\ e^{i\varphi} = 0. \end{bmatrix} \tag{1.56a}$$

In this case, the equation $e^{i\varphi} = 0$ cannot be omitted, because neglecting a root of this equation, we lose a part of the set of singularities (which is essential for some figures).

But the singularities of the function $U^0(\vec{r})|_S$ are located at the points of the "image" of the initial field sources at which they arrive along the complex characteristics. The coordinates of these images can be determined by using the Riemann-Schwartz symmetry principle [10,49].

Now we consider another method for determining the coordinates of the point source image. Let

$$U^0(\vec{r}) = \frac{1}{4i}H_0^{(2)}(k|\vec{r} - \vec{r}_0|)$$

be the point source field. We have

$$|\vec{r} - \vec{r}_0|^2\big|_S = \rho^2(\varphi) + r_0^2 - 2r_0\rho(\varphi)\cos(\varphi - \varphi_0)$$
$$= [\rho(\varphi)e^{i\varphi} - r_0e^{i\varphi_0}][\rho(\varphi)e^{-i\varphi} - r_0e^{-i\varphi_0}].$$

Thus, the function $U^0(\vec{r})|_S$ has singularities at

$$\rho(\varphi)e^{\pm i\varphi} = r_0e^{\pm i\varphi_0}.$$

Moreover, the function $U^0(\vec{r})$ has a singularity at infinity.

The root of the equation $\rho(\varphi)e^{i\varphi} = r_0e^{i\varphi_0}$ corresponds to the singularity of the function $U^0(\vec{r})$ at the source point $r_0e^{i\varphi_0} = x_0 + iy_0$. Now we consider the equation

$$\rho(\hat{\varphi})e^{-i\hat{\varphi}} = r_0e^{-i\varphi_0} \equiv z^-. \tag{1.57}$$

If we assume that $\hat{\varphi}$ is real in this equation, then we obtain the same point as in the preceding case. Therefore, we solve Equation (1.57) under the assumption that $\hat{\varphi} = \varphi_1 + i\varphi_2$ is complex. Formulas (1.55) and (1.57) imply

$$\zeta_{\text{sing}}\exp(-i2\hat{\varphi}(z^-)) = z^-,$$

i.e.,

$$\zeta_{\text{sing}} = z^-\exp(i2\hat{\varphi}(z^-)), \tag{1.58}$$

where $\zeta_{\text{sing}} \equiv \zeta(\hat{\varphi}(z^-))$.

Precisely relation (1.58) determines the coordinates of the source "image." In addition, as already noted, it is also necessary to find the image of the infinitely remote point.

So we obtain the relations

$$\sigma_0 = \max_{\varphi_0}\left|\frac{k\rho(\varphi_0)}{2}\exp(i\varphi_0)\right|, \tag{1.59}$$

$$\sigma_{S0} = \frac{k}{2}\max_{\varphi_0}\text{Im}\{\rho(\varphi_0)\exp i\varphi_0\}, \tag{1.60}$$

where φ_0 are roots of Equation (1.56a).

Formulas (1.59) and (1.60) hold if $U^0(\vec{r})$ is the plane wave field. However, in the general case, when determining the value of σ, it is also necessary to consider the singular points determined by formula (1.58). Thus, in the general case, we have

$$\sigma = \max\left(\sigma_0, \frac{k}{2}|\zeta_{\text{sing}}|\right).$$

Similarly, for σ_S, we have

$$\sigma_S = \max\left(\sigma_{S0}, \frac{k}{2}\text{Im}\zeta_{\text{sing}}\right).$$

If σ_S is known for all orientations of the coordinate system, then the convex envelope $\bar{B}$ of singularities of the wave field can be determined [9, Chapter 2].

The above technique for localizing the singular points can be transferred to the three-dimensional case without any changes in the case, where the scatterer is a body of revolution. Of course, it is natural that relations (1.56) and (1.58) allow us to determine the coordinates of singular points in the axial cross-section of the scatterer.

However, in the general situation in the three-dimensional case, the scatterer boundary S can analytically be deformed if we introduce the complex variable

$$\zeta = \rho(\theta, \varphi)e^{i\theta} \tag{1.61}$$

(where $r = \rho(\theta, \varphi)$ is the equation of the boundary S in spherical coordinates) and assume that the angle θ is a complex variable and the angle φ is a parameter. As a result, proceeding by analogy with the two-dimensional case, we seek the coordinates of the singularities by solving the equations

$$\left(\rho'_\theta(\theta, \varphi) + i\rho(\theta, \varphi)\right)\exp(i\theta)\big|_{\theta_0, \varphi_0} = 0,$$

$$\rho'_\varphi(\theta, \varphi)\big|_{\theta_0, \varphi_0} = 0. \tag{1.62}$$

Now we have [10,41]

$$\sigma_0 = \max_{\theta_0, \varphi_0}\left|\frac{k\rho(\theta_0, \varphi_0)}{2}\exp(i\theta_0)\right|, \tag{1.63}$$

$$\sigma_{S0} = \max_{\theta_0, \varphi_0}\text{Im}\left\{\frac{k\rho(\theta_0, \varphi_0)}{2}\exp(i\theta_0)\right\}. \tag{1.64}$$

To conclude this section, we consider the problem of localization of the singularities in the case of plane wave scattering at an infinite periodic surface $y = h(x) = h(x + b)$, where b is the surface period [45,50]. As previously, this can be performed using representation (1.9) with (1.12). We again use the technique of analytic deformation of the boundary. By analogy with (1.55), we introduce the complex variable (here it is more convenient to use the Cartesian coordinates)

$$\zeta = x + iy(x). \tag{1.65}$$

On the complex plane $z = x + iy$, relation (1.65) associates the real values of the variable x with the line $y = h(x)$. But if we assume that the variable x in (1.65) is complex and set $\hat{x} = x_1 + ix_2$, then the curve $y = h(x)$ begins to deform continuously until the mapping (1.65) ceases to be one-to-one. The points at which the mapping (1.65) is not one-to-one can be determined by the equation

$$\zeta'(x) \equiv 1 + iy'(x) = 0. \tag{1.66}$$

We denote the roots of Equation (1.66) by x_0 and then have the singularities on the complex plane $z = x + iy$ at the points $\zeta(x_0)$. Separating the real and imaginary parts of $\zeta(x_0)$, we determine the coordinates of the singular points of the analytic continuation of the wave field to the half-plane beyond the interface $y = h(x)$.

Thus, in particular (see (1.40) and (1.41)) [9,45], we have

$$\sigma_S = \frac{k}{2} \max_{x_0} \mathrm{Im}\zeta_0. \tag{1.67}$$

Several examples of singularity localization are considered below.

1.2.3 Examples of Determining the Singularities of the Wave Field Analytic Continuation

1.2.3.1 Singularities of Mapping (1.55)

These singularities can be determined by solving Equations (1.56a). Let us consider several examples of solving these equations for different geometries.

1. *Elliptic cylinder.* As the first example, we consider the problem of plane wave diffraction at an elliptic cylinder. In this case, the equation of the cross-section contour S has the form

$$\rho(\varphi) = \frac{b}{\sqrt{1 - \varepsilon^2 \cos^2\varphi}},$$

where b is the minor semiaxis, $a = \frac{b}{\sqrt{1-\varepsilon^2}}$ is the major semiaxis, and $\varepsilon = \sqrt{1 - \frac{b^2}{a^2}}$ is the eccentricity.

After elementary transformations, we obtain

$$\cos^2\varphi_0 - i\cos\varphi_0\sin\varphi_0 - 1/\varepsilon^2 = 0$$

for the complex root φ_0 of the first equation in (1.56a), which implies

$$e^{i\varphi_0} = \pm\frac{\varepsilon}{\sqrt{2-\varepsilon^2}},$$

that is,

$$\zeta_{01} \equiv \zeta(\varphi_0) = \pm\frac{b\varepsilon}{\sqrt{1-\varepsilon^2}} = \pm a\varepsilon = \pm f,$$

where $2f$ is the interfocus distance. Thus, we see that the wave field singularities inside the ellipse lie at its foci. The second equation in (1.56a) is associated with the singular point at the origin. Equation (1.56) has no roots outside the ellipse, which means that the solution of the internal boundary-value problem for the domain inside the ellipse can be continued everywhere into the outer domain. One can show (see below) that the wave field has second-order branching points at the ellipse foci. To distinguish a single-valued branch, it is necessary to connect the points $\pm f$ by a cut. For σ and σ_S, we respectively have (see Equations (1.59) and (1.60))

$$\sigma = \frac{kf}{2}, \quad \sigma_S = 0.$$

In the three-dimensional case, where the scatterer is a spheroid, the singularities lie at the foci of its axial cross-section.

2. *Cassinian oval.* As the second example, we consider the case, where the shape of the boundary S is the Cassinian oval:

$$\rho(\varphi) = a\sqrt{1 + \varepsilon^2\cos^2\varphi}, \quad \varepsilon < 1.$$

The first equation in (1.56a) can be rewritten as

$$\varepsilon^2\cos^2\varphi_0 + i\varepsilon^2\cos\varphi_0\sin\varphi_0 + 1 = 0.$$

Solving this equation, we obtain

$$e^{i\varphi_0} = \pm i\frac{\sqrt{2+\varepsilon^2}}{\varepsilon}$$

and hence we have

$$\zeta_{02} = \pm i\frac{a}{\varepsilon}\sqrt{1+\varepsilon^2},$$

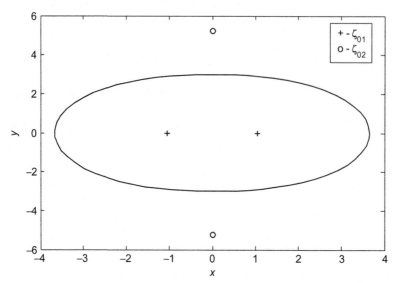

Figure 1.1 Cassinian oval.

that is, $|\zeta_{02}| = \frac{a}{\varepsilon}\sqrt{1+\varepsilon^2} > \rho\left(\pm\frac{\pi}{2}\right) = a$, $\arg\zeta_{02} = \pm\frac{\pi}{2}$. Therefore, the points ζ_{02} lie outside S (see Fig. 1.1).
Solving the second equation in (1.56a), we obtain

$$\zeta(\varphi_0) = \pm\frac{a\varepsilon}{2} = \zeta_{01}.$$

These singular point lies inside S, because $|\zeta_{01}| = \frac{a\varepsilon}{2} < \rho(0) = \rho(\pi) = a\sqrt{1+\varepsilon^2}$.
For the Cassinian oval, we thus have

$$\sigma = \frac{ka\varepsilon}{4}, \quad \sigma_S = 0.$$

3. *Multifoil.* Now we consider the case, where the boundary S is a multifoil whose equation has the form

$$\rho(\varphi) = a(1 + \tau\cos q\varphi), \quad 0 \le \tau < 1, q = 1, 2, 3, \ldots \qquad (1.68)$$

3a. The case $q = 1$ should be considered separately. In this case, the curve (1.68) is called the Pascal limacon. As in the preceding example, it is here necessary to consider both equations in (1.56a). Solving the second equation, we obtain

$$\zeta(\varphi_0) = \frac{a\tau}{2} = \zeta_{01}.$$

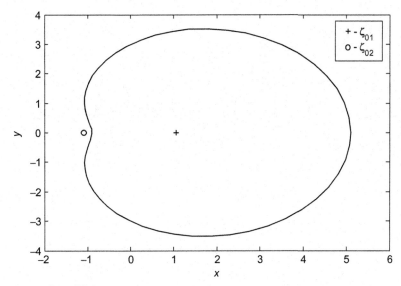

Figure 1.2 Pascal limacon.

It is easy to see that this point lies inside S (see Fig. 1.2). In this example, the first equation in (1.56a) has the form

$$\rho'(\varphi_0) + i\rho(\varphi_0) = -a\tau\sin\varphi_0 + ia + ia\tau\cos\varphi_0 = 0$$

and the solution $e^{i\varphi_0} = -1/\tau$, that is,

$$\zeta_{02} \equiv \rho(\varphi_0)e^{i\varphi_0} = a\left[1 + \frac{\tau}{2}\left(e^{i\varphi_0} + e^{-i\varphi_0}\right)\right]e^{i\varphi_0} = -\frac{a}{2\tau}\left(1 - \tau^2\right)$$

Thus, we have $|\zeta_{02}| = \frac{a}{2\tau}(1 - \tau^2)$, $\arg\zeta_{02} = \pi$, and therefore, the point ζ_{02} lies outside S (see Fig. 1.2).

   3b. Now we consider the cases, where $q = 2,3,\ldots$. Here the first equation in (1.56a) becomes

$$\rho'(\varphi_0) + i\rho(\varphi_0) = a(-q\tau\sin q\varphi_0 + i + i\tau\cos q\varphi_0) = 0.$$

We denote $e^{i\varphi_0} = t$ and obtain the set of solutions [9, Chapter 2]

$$t_{1m} = \left[\frac{-1 + \sqrt{1 + \tau^2(q^2 - 1)}}{\tau(q+1)}\right]^{1/q} \exp\left(i\frac{2m\pi}{q}\right), \quad m = 0, 1, \ldots, q-1, \quad (1.69)$$

$$t_{2m} = \left[\frac{1 + \sqrt{1 + \tau^2(q^2 - 1)}}{\tau(q+1)}\right]^{1/q} \exp\left(i\frac{(2m+1)\pi}{q}\right), \quad m = 0, 1, \ldots, q-1.$$

$$(1.70)$$

Now because $\zeta = \rho(\varphi)e^{i\varphi}$ (see Equation (1.55)), we use the obtained solutions of Equations (1.69) and (1.70) to show that the singularities of the wave field analytic continuation into the interior of the multifoil must be located at points lying at the distance

$$|\zeta_{01}| = a\frac{q[q + \sqrt{1+\tau^2(q^2-1)}]}{q^2-1}\left[\frac{-1+\sqrt{1+\tau^2(q^2-1)}}{\tau(q+1)}\right]^{1/q} \tag{1.71}$$

from the origin on the lays drawn at the angles

$$\arg\zeta_{01} = \frac{2m\pi}{q}, \quad m = 0, 1, \ldots, q-1. \tag{1.72}$$

Similarly, the singularities of the wave field continuation to the outer domain are at the points at the distance

$$|\zeta_{02}| = a\frac{q[q - \sqrt{1+\tau^2(q^2-1)}]}{q^2-1}\left[\frac{1+\sqrt{1+\tau^2(q^2-1)}}{\tau(q+1)}\right]^{1/q} \tag{1.71a}$$

from the origin on the rays drawn at the angles

$$\arg\zeta_{02} = \frac{(2m+1)\pi}{q}, \quad m = 0, 1, \ldots, q-1 \tag{1.72a}$$

(see Fig. 1.3). The roots of the equation $e^{i\varphi} = 0$ are mapped into the infinitely remote point on the plane z.

It follows from Equations (1.59) and (1.71) that the pattern degree σ for multifoils is equal to

$$\sigma = \frac{kaq[q + \sqrt{1+\tau^2(q^2-1)}]}{2\quad q^2-1}\left[\frac{-1+\sqrt{1+\tau^2(q^2-1)}}{\tau(q+1)}\right]^{1/q}, \quad q = 2, 3, \ldots.$$

We note that the field has, for example, algebraic branching of the qth order at singular points (see Equations (1.71) and (1.72)) inside the scatterer [9, Chapter 2], [48]. More precisely, the singularity character near the mth singular point $(m = 0, 1, \ldots, q-1)$ has the form (see Section 1.2.5):

$$\frac{1}{(re^{i\varphi} - r_{0m}e^{i\varphi_{0m}})^{1-1/q}}, \tag{1.73}$$

where r, φ are polar coordinates of a point near the mth $(m = 0, 1, \ldots, q-1)$ singular point with coordinates $r_{0m} = |\zeta_{01}|$, $\varphi_{0m} = \arg\zeta_{01}$. To distinguish

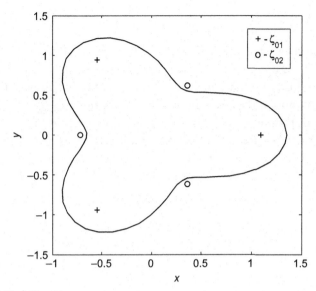

Figure 1.3 Trefoil.

a unique branch, it is necessary to draw cuts from these points to the origin, and from the points corresponding to the roots (Equation (1.70)), to the infinitely remote point (see Fig. 1.3).

In particular, it follows from Equation (1.73) that the singularity character becomes similar to that of a singularity at a first-order pole as $q \to \infty$. One can also see from Equation (1.71) that the singular points tend to the origin as $\tau \to 0$. Thus, if we consider a circle (to a sphere in the three-dimensional case), that is, if we simultaneously let $q \to \infty$ and $\tau \to 0$, then we obtain one singular point at the origin and this is a pole of an infinite order, that is, an essentially singular point.

In other words, the field scattered at a circular cylinder (sphere) can be represented either by sources occupying an infinitely small volume and containing the origin (also see Equation (1.2) in view of the above argument) or by a single source, that is, a multipole of an infinite order at the origin. The following question arises: Why do we still see, for example, a sphere rather than a luminous point at its center? An attempt to answer this question can be made only after our consideration of the second set of singularities, that is, the images of the external sources (singular points).

The singularities of the analytic continuation of the wave field scattered at an obstacle whose cross-section is one of the canonical second-order curves (i.e., ellipse, hyperbola, and parabola) can be localized uniquely if we use the equation of these curves in polar coordinates with origin at one of the foci of the corresponding curve. As is known, the equation has the form $\rho(\varphi) = \frac{p}{1-\varepsilon\cos\varphi}$, where p is a (focal) parameter and ε is the eccentricity, which is $0 \le \varepsilon < 1$ for the ellipse, $\varepsilon > 1$ for the hyperbola, and $\varepsilon = 1$ for the parabola.

According to Equation (1.55), the second equation in (1.56a) then leads to a singularity at the point

$$\zeta(\varphi_0) \equiv \rho(\varphi_0)e^{i\varphi_0} = 0,$$

that is, at the focus considered as the origin of polar coordinates. In the case under study, the first equation in (1.56a) becomes

$$\varepsilon\sin\varphi_0 - i + i\varepsilon\cos\varphi_0 = 0$$

and has the solution $e^{i\varphi_0} = \varepsilon$.

As a result, we obtain

$$\zeta(\varphi_0) = \frac{2p\varepsilon}{1-\varepsilon^2}.$$

In the case of ellipse, $p = \frac{b^2}{a}$, that is, $\zeta(\varphi_0) = 2a\varepsilon = 2f$. This means that the second singular point lies at the second (right) focus, as was shown above.

In the case of hyperbola, where $\varepsilon = \frac{c}{a} > 1$, a is the real semiaxis of the hyperbola, and $2c$ is the distance between the foci, we have $\zeta(\varphi_0) = -2c$, which corresponds to the singularity at the second (left) focus of the hyperbola.

Finally, in the case of parabola, the root of the first equation in (1.56a) corresponds to the singularity at the infinitely remote point, that is, in the case of parabolic boundary, there is only one singular point in the finite region of the space, that is, the singularity at the parabola focus (also see below).

4. *Three-axial ellipsoid.* Now we consider the three-dimensional problem. We assume that the scatterer boundary S is a three-axial ellipsoid

$$\frac{x^2}{a^2} + \frac{y^2}{b^2} + \frac{z^2}{c^2} = 1,$$

or in polar coordinates

$$\rho(\theta, \varphi) = \frac{c}{\sqrt{\dfrac{\sin^2\theta}{\kappa^2} + \cos^2\theta}} = c\sqrt{\dfrac{1 + tg^2\theta}{1 + \dfrac{1}{\kappa^2}tg^2\theta}},$$

where $\frac{1}{\kappa^2} = \frac{c^2}{a^2 b^2}(a^2\sin^2\varphi + b^2\cos^2\varphi)$.

It follows from the second equation in (1.62) that $\varphi_0 = m(\pi/2)$, $m = 0, 1, 2, 3$. However, the first equation can be reduced to the form

$$tg^2\theta + i(\kappa^2 - 1)tg\theta + \kappa^2 = 0.$$

This equation has the solution $e^{i\theta_0} = \sqrt{\dfrac{1 + \kappa_0^2}{1 - \kappa_0^2}}$. Now $\rho(\theta_0, \varphi_0) = c\sqrt{\dfrac{1 - b^2}{1 - (b^2/\kappa_0^2)}}$, $\kappa_0^2(0) = \frac{a^2}{c^2}$, and $\kappa_0^2(\pi/2) = \frac{b^2}{c^2}$. The quantities σ and σ_S can be obtained from the relations in Equations (1.63) and (1.64), respectively, and are determined by the relations between the ellipsoid semiaxes. For example, $\sigma_S = \sigma = \frac{k}{2}\sqrt{c^2 - a^2}$ for $c > b > a$.

5. *Periodic surface.* Now we consider the problem of diffraction at a periodic surface $y = h(x) = h(x + b)$. As the first example, we consider the surface $h(x) = a\cos px$, $a > 0, p = 2\pi/b$ [9,50]. In this case, Equation (1.66) has the form

$$1 - iap \sin px = 0$$

and the solutions

$$x_0^1 = -\frac{i}{p}\ln\frac{1 + \sqrt{1 + a^2 p^2}}{ap} + 2q\frac{\pi}{p}, \quad q = 0, \pm 1, \ldots,$$

$$x_0^2 = \frac{i}{p}\ln\frac{1 + \sqrt{1 + a^2 p^2}}{ap} + (2q + 1)\frac{\pi}{p}, \quad q = 0, \pm 1, \ldots.$$

We substitute the obtained solutions in Equation (1.65) and determine the coordinates of the singular points:

$$\zeta_0^1 = -ia\left(\frac{1}{ap}\ln\frac{1 + \sqrt{1 + a^2 p^2}}{ap} - \frac{\sqrt{1 + a^2 p^2}}{ap}\right) + qb, \quad q = 0, \pm 1, \ldots,$$

$$\zeta_0^2 = ia\left(\frac{1}{ap}\ln\frac{1 + \sqrt{1 + a^2 p^2}}{ap} - \frac{\sqrt{1 + a^2 p^2}}{ap}\right) + (2q - 1)\frac{b}{2}, \quad q = 0, \pm 1, \ldots.$$

$$\text{(1.74)}$$

The singular points ζ_0^1 arise in the analytic continuation across the interface $y = a\cos px$ of the plane wave field scattered at this surface. Similarly, the points ζ_0^2 are singularities of the field transmitted into the lower half-space (under the surface $y = a\cos px$) when it is analytically continued into the upper half-space.

Thus, in particular (see Equation (1.67)),

$$\sigma_S = \frac{ka}{2}\left(\frac{\sqrt{1 + a^2 p^2}}{ap} - \frac{1}{ap}\ln\frac{1 + \sqrt{1 + a^2 p^2}}{ap}\right).$$

Now we consider the second example, that is, the plane wave scattering at a cycloid [51] parametrically determined by the equations

$$x = a(t + \tau\cos t), \quad y = a\tau\sin t, \quad a > 0, 0 < \tau \leq 1.$$

Here we set

$$\zeta \equiv x + iy = at + a\tau e^{it}.$$

Further, we have

$$\zeta_t' = a + ia\tau e^{it} = 0.$$

Determining the roots of this equation

$$t_0 = \frac{\pi}{2} + 2\pi q + i\ln\tau, \quad q = 0, \pm 1, \ldots,$$

we obtain

$$\zeta_0 = a\left(\frac{\pi}{2} + 2\pi q\right) + ia(1 + \ln\tau), \quad q = 0, \pm 1, \ldots.$$

No singularities arise when the field transmitted into the half-space under the cycloid is analytically continued to the upper half-space [51]. Here we have

$$\sigma_S = \frac{ka}{2}(1 + \ln\tau).$$

1.2.3.2 Singularities at Source Images

The coordinates of these singularities can be determined from the relations (1.57) and (1.58). We again consider several examples of their localization for different scatterer geometries.

1. *Ellipse.* Here Equation (1.57) becomes

$$\frac{be^{-i\hat{\varphi}}}{\sqrt{1-\varepsilon^2\cos^2\hat{\varphi}}} = r_0 e^{-i\varphi_0} \equiv z^-,$$

that is,

$$4b^2 e^{-i2\hat{\varphi}} = (z^-)^2 \left[4 - \varepsilon^2 \left(e^{i2\hat{\varphi}} + e^{-i2\hat{\varphi}} + 2\right)\right].$$

We denote $e^{-i2\hat{\varphi}} = x$. Then we have

$$x^2 - 2\frac{(2-\varepsilon^2)}{\varepsilon^2 + 4\dfrac{b^2}{(z^-)^2}} x + \frac{\varepsilon^2}{\varepsilon^2 + 4\dfrac{b^2}{(z^-)^2}} = 0,$$

which implies

$$x_{1,2} = \frac{(2-\varepsilon^2) \pm 2\sqrt{\dfrac{b^2}{(z^-)^2} - \dfrac{b^2\varepsilon^2}{(z^-)^2}}}{\varepsilon^2 + 4\dfrac{b^2}{(z^-)^2}}.$$

Thus, we finally have

$$e^{-i2\hat{\varphi}} = \frac{(2-\varepsilon^2) + 2\sqrt{\dfrac{b^2}{(z^-)^2} - \dfrac{b^2}{r_0^2}\varepsilon^2 e^{i2\varphi_0}}}{\varepsilon^2 + 4\dfrac{b^2}{r_0^2}e^{i2\varphi_0}}.$$

The root x_2 is redundant, because it gives a false result for $\varepsilon = 0$. So we have

$$\zeta_{sing} = r_0 e^{-i\varphi_0} e^{i2\hat{\varphi}} = r_0 \frac{\varepsilon^2 e^{-i\varphi_0} + \left(\dfrac{2b}{r_0}\right)^2 e^{i\varphi_0}}{(2-\varepsilon^2) + 2\sqrt{\dfrac{b^2}{a^2} - \dfrac{b^2}{r_0^2}\varepsilon^2 e^{i2\varphi_0}}}. \tag{1.75}$$

It follows from Equation (1.75) that the infinitely remote point maps into to the origin.

1a. Special case $\varepsilon = 0$ (a circle or a sphere). In this case, it follows from the relation in Equation (1.75) that

$$\zeta_{sing} = \frac{b^2}{r_0} e^{i\varphi_0}.$$

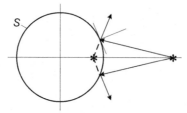

Figure 1.4 Image in a circle.

that is, the image is at the inversion point when the product of the distances from the circle (sphere) center to the source and to the image is equal to the squared radius of this circle (see Fig. 1.4).

1b. Let $r_0 = b + \delta$, $\varphi_0 = \pi/2$, $\delta \ll b$, that is, let the source lie near the boundary. Preserving the quantities of the order of δ and omitting the quantities of the order of δ^2, we obtain

$$\zeta_{\text{sing}} \approx ir_0 \frac{4\left(1 - 2\dfrac{\delta}{b}\right) - \varepsilon^2}{(2 - \varepsilon^2) + 2\sqrt{\left(\dfrac{b}{a}\right)^2 + \varepsilon^2 \left(1 - 2\dfrac{\delta}{b}\right)}} \approx ir_0 \frac{4 - \varepsilon^2 - 8\dfrac{\delta}{b}}{4 - \varepsilon^2 - 2\varepsilon^2 \dfrac{\delta}{b}}$$

$$\approx i(r_0 - 2\delta).$$

Thus, approximately (i.e., up to quantities of the order of $(\delta/b)^2$), the image is at the "mirror" point, as this must be according to the visual experience. As the source moves away along the line $\varphi_0 = \pi/2$, Equation (1.75) shows that the image approaches the segment connecting the foci until it arrives at the point $r = 0$ on the cut for $r_0 = \frac{2a}{\varepsilon}$. As r_0 continues to increase, the source image intersects the cut (because it appears for $\varphi_0 = -\pi/2$), that is, it appears on the non-physical leave of the Riemannian surface. Thus, for $r_0 > \frac{2a}{\varepsilon}$, the set of the diffraction field singularities is only the interfocus segment, as in the case of an ellipse excitation by a plane wave.

It should be noted that the relations in Equations (1.57) and (1.58) allow us to determine not only the coordinates of the point source image but also the coordinates of singular points generated by the wave field singularities located inside a neighboring body, for example, in the situation illustrated in Fig. 1.5.

The information about such singular points can be used to solve the diffraction problems on a group of bodies [52].

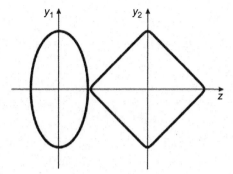

Figure 1.5 Superellipse near an ellipse.

2. *Pascal limacon.* We recall that its equation has the form $\rho(\varphi) = a + b\cos\varphi$, $0 \le b < a$. Here Equation (1.57) becomes

$$(a + b\cos\hat{\varphi})e^{-i\hat{\varphi}} = r_0 e^{-i\varphi_0} \equiv z^-,$$

or

$$e^{-i2\hat{\varphi}} + \frac{2a}{b}e^{-i\hat{\varphi}} + \left(1 - \frac{2z^-}{b}\right) = 0,$$

which implies

$$e^{-i\hat{\varphi}} = \frac{-a \pm \sqrt{a^2 - b^2 + 2bz^-}}{b},$$

that is,

$$e^{i\hat{\varphi}} = \frac{b}{-a + a\sqrt{1 - \dfrac{b^2 + 2bz^-}{a^2}}}.$$

Here we omit the solution with the sign "−" of the root, because this solution does not become a solution for the circle for $b = 0$: $e^{i\hat{\varphi}} = \frac{a}{r_0}e^{i\varphi_0}$. Therefore, we have

$$\zeta_{\text{sing}} \equiv r_0 e^{-i\varphi_0} e^{i2\hat{\varphi}} = \frac{r_0 b^2 e^{-i\varphi_0}}{2a^2 - b^2 + 2r_0 b e^{-i\varphi_0} - 2a\sqrt{a^2 - b^2 + 2r_0 b e^{-i\varphi_0}}}.$$

Let us verify the obtained solution. Let $r_0 = a + b + \delta$, $\varphi_0 = 0$, $\delta \ll a + b$. We again preserve the quantities of the order of δ and neglect the quantities of the order of δ^2 to obtain

$$\zeta_{\text{sing}} \approx \frac{b^2(a+b+\delta)}{a^2+(a+b)^2+2b\delta-2a(a+b)\left(1+\dfrac{b\delta}{(a+b)^2}\right)} \approx (a+b)\frac{(a+b+\delta)}{(a+b+2\delta)},$$

or finally,

$$\zeta_{\text{sing}} \approx a+b-\delta,$$

that is, as in the case of ellipse, the image lies at the mirror point.

It follows from the above examples that the image of a point source (singular point) can be obtained by the method of expansion in a small parameter.

3. *Parabola.* As the last example, we consider the problem of excitation of a mirror shaped as a parabolic cylinder by a point source (more precisely, by a filament of current) $U^0(\vec{r}) = \frac{1}{4i}H_0^{(2)}(k|\vec{r}-\vec{r}_0|)$ [9, Chapter 2]. This problem is of interest in applications, in particular, in the theory of antennas.

Therefore, we assume that the point source field is incident on the parabola

$$y = a - bx^2. \tag{1.76}$$

As in the previous cases, we introduce the complex variable $\zeta = x + iy(x)$.

In the case of parabola (1.76), Equation (1.66) becomes $1 - i2bx = 0$ and has the solution $x_0 = (1/2bi)$.

On the complex plane z, this solution is associated with the singular point

$$\zeta_0 = i\left(a - \frac{1}{4b}\right).$$

It is necessary to draw a cut from this point to the infinitely remote point. On the plane of real coordinates (x, y), the singular point ζ_0 is associated with the point with the coordinates

$$x_0 = 0, \quad y_0 = a - \frac{1}{4b}, \tag{1.77}$$

which is the parabola focus.

Now let us determine the point source images. In this case, we proceed precisely as above.

Therefore, we again consider the expression (now in Cartesian coordinates)

$$|\vec{r} - \vec{r}_0|^2\big|_S = [(x + iy(x)) - (x_0 + iy_0)][(x - iy(x)) - (x_0 - iy_0)].$$

Thus, the function $U^0(\vec{r})$ has singularities for

$$(\hat{x} \pm iy(\hat{x})) = (x_0 \pm iy_0) \equiv z_0^{\pm}.$$

On the plane $z = x + iy$, the equation $\zeta \equiv \hat{x} + iy(\hat{x}) = z_0^{+}$ is associated with the point $z_0 = x_0 + iy_0$, which is not interesting for us, because this is the point of the source location. Therefore, we consider the point at which

$$\hat{x} - iy(\hat{x}) = z_0^{-}, \tag{1.78}$$

where $\hat{x} = x_1 + ix_2$ is a complex quantity.

Relation (1.78) is equivalent to

$$2\hat{x}(\zeta_{\text{sing}}) - \zeta_{\text{sing}} = z_0^{-}, \tag{1.79}$$

where $\hat{x}(\zeta)$ is a solution of the equation

$$\hat{x} + iy(\hat{x}) = \zeta, \tag{1.80}$$

which is defined everywhere except the points where formula (1.66) is satisfied.

We rewrite Equation (1.78) as

$$\hat{x} - ia + ib\hat{x}^2 = z_0^{-}.$$

Its solution is

$$\hat{x}(z_0^{-}) = \frac{i \pm i\sqrt{1 - 4ab + 4ibz_0^{-}}}{2b},$$

and the solution with the upper sign is redundant, because it is false for $b = 0$. Therefore, we have (see (1.79)),

$$
\begin{aligned}
\zeta_{\text{sing}} &\equiv 2\hat{x}(z_0^{-}) - z_0^{-} = \frac{i - i\sqrt{1 - 4ab + 4ibz_0^{-}}}{b} - z_0^{-} \\
&= \frac{i - b(x_0 - iy_0) - i\sqrt{1 - 4ab + 4ib(x_0 - iy_0)}}{b}.
\end{aligned} \tag{1.81}
$$

Separating the real and imaginary parts of (1.81), we obtain the coordinates of the point source image in a parabolic mirror.

Now let us consider the limit cases [9, Chapter 2]. First, we assume that $|ab| \ll 1$, $|bz_0| \ll 1$ (i.e., we consider the case of a weakly curved surface). Then, as follows from (1.81), $\zeta_{\text{sing}} = (x_0 - iy_0) + i2a$. Thus, the source image on the plane (x, y) is located at the point with the coordinates

$$x_{\text{sing}} = x_0, \quad y_{\text{sing}} = -y_0 + 2a, \tag{1.82}$$

i.e., it is at the same point, where it was when reflected from the plane $y = a$. Similarly, for $|bx_0| \ll 1$, $|by_0| \ll 1$ (i.e., in the case where the source is located near the surface), the image is again at the point with the coordinates (1.82).

Comparing the obtained result with the results for the compact scatterers considered above, we can formulate the following locality principle: *if a source is located near a convex curvilinear surface at a distance which is small compared with the curvature radius of this surface at the point under study, then its image is (approximately) located as in the case of reflection from the tangent plane.*

Now we can try to answer the question posed above: "Why do we see a sphere rather than a luminous point at its center?" Obviously, it is impossible to answer this question completely only within the framework of the diffraction theory. However, it follows from the above examples that if we place an ideally polished sphere (of course, on the scales exceeding the eye resolution) in the plane wave field (e.g., irradiating the sphere by a laser ray) in a room with absorbing walls, we indeed do not see anything except a luminous point at the sphere's center. The brightness of this point is determined by the properties of the sphere's surface; for example, an ideally polished sphere made of an absorbing material looks like a dim point in this situation. In the usual situation, we see, for example, the same sphere surrounded by many other objects whose images on this sphere form the visual image that we perceive as a sphere. However, if the sphere's surface is rough, this actually means that we deal with an object whose shape is nearly spherical with a set of singular points on the surface and (or) near it, and precisely these points are sources of the scattered field.

Generalizing the above reasoning to the case of nonspherical objects allows us get a basic understanding of "why we see what we see". Indeed, the sources of the field scattered at an object with mirror surface are located inside the object. Therefore, such an object is easier to recognize the more foreign objects are located near it and reflected at it. Conversely, if an object has a rough surface, then the scattered field sources lie directly on this surface, which makes the problem of the object shape recognition very simple.

Thus, we can conclude that for a scatter to have the properties of an "invisible being", i.e., to be badly recognizable, its shape must be as close as possible to an ideal analytic surface without angles, protuberances, and other irregularities, and its surface must have absorbing properties [10]. Here the following should be noted. It is obvious that any mirror surface becomes irregular at a sufficient magnification. Thus, the degree of the surface roughness is determined not only by the surface properties but also by the wavelength of the incident field and by the device resolution or the sensitivity of the organ perceiving the image. The same is true for the definition of analytic surfaces.

1.2.4 Boundaries of the Domains of Existence of Analytic Representations

Constructing the solution of a boundary-value problem for the Helmholtz equation (system of Maxwell equations) often starts from choosing an analytic representation of the solution, for example, the expansion in a series in a complete system of functions where each term separately satisfies the equation. The coefficients of this expansion are then sought by using the boundary conditions. In this case, it is very important that the domain of existence of such a representation contains the domain where the solution of the boundary-value problem is sought, because otherwise, the corresponding numerical algorithms are inadequate to the initial boundary-value problem and ill conditioned.

The exact boundaries of the domains of existence of different analytic representations of solutions to the Helmholtz equation can be obtained by using the relations obtained above.

First, we consider the wave field representation by a series in wave harmonics (Rayleigh series) (1.29), (1.33). Let consider the two-dimensional case in more detail [9, Chapter 2]. We need a relation that allows us to express the coefficients a_m of this series in terms of the wave field values on the scatterer boundary S. This relation can easily be obtained from representation (1.9) for the wave field. For simplicity, we assume that the Dirichlet condition is satisfied on S (the results are independent of the form of the boundary conditions). Then the integral (1.9) can be written in polar coordinates as

$$U^1(r, \varphi) = \int_0^{2\pi} q(\varphi') H_0^{(2)} \left(k\sqrt{r^2 + \rho^2(\varphi') - 2r\rho(\varphi')\cos(\varphi - \varphi')} \right) d\varphi',$$

$$(1.83)$$

where $r = \rho(\varphi')$ is the equation of the scatterer boundary S and

$$q(\varphi) = \frac{i}{4}\left[\rho(\varphi)\frac{\partial U}{\partial r} - \frac{\rho'(\varphi)}{\rho(\varphi)}\frac{\partial U}{\partial \varphi}\right]_{r=\rho(\varphi)}. \qquad (1.84)$$

The following expansion holds for all $r > R_0$ (where R_0 is the radius of the circle circumscribed about S) [40]:

$$H_0^{(2)}\left(k\left|\vec{r} - \vec{r}'\right|\right) = \sum_{n=-\infty}^{\infty} J_n(k\rho(\varphi'))H_n^{(2)}(kr)e^{in(\varphi - \varphi')}. \qquad (1.85)$$

Now we obtain representation (1.29) from (1.83) and (1.85), where

$$a_m = i^m \int_0^{2\pi} q(\varphi)J_m(k\rho(\varphi))e^{-im\varphi}d\varphi. \qquad (1.86)$$

Since

$$J_m(k\rho) = \frac{1}{|m|!}\left(\frac{k\rho}{2}\frac{m}{|m|}\right)^{|m|}\left(1 + O\left(\frac{1}{m}\right)\right)$$

as $|m| \to \infty$ [35,40], we asymptotically have

$$a_m = \frac{i^{|m|}}{|m|!}\int_0^{2\pi} q(\varphi)\left(\frac{k\rho(\varphi)}{2}\right)^{|m|}e^{-im\varphi}d\varphi\left(1 + O\left(\frac{1}{m}\right)\right). \qquad (1.87)$$

By analogy with the above consideration (see Section 1.2.1 and, in particular, relation (1.47)), we can write the following representation for the function $f^E(w_\pm)$:

$$f^E(w_\pm) = \int_0^{2\pi} q(\varphi)\exp\left(\frac{ik\rho(\varphi)}{2}w_\pm e^{is_\beta\varphi}\right)d\varphi. \qquad (1.88)$$

From (1.88), using the expansion of the exponential function in a power series, we obtain the coefficients c_m in expansion (1.53):

$$c_m = \frac{i^m}{m!}\int_0^{2\pi} q(\varphi)\left(\frac{k\rho(\varphi)}{2}\right)^m e^{ims_\beta\varphi}d\varphi, \quad m \geq 0. \qquad (1.89)$$

Comparing formulas (1.87) and (1.89), we obtain the asymptotic relation

$$a_{|m|} = c_m\left(1 + O\left(m^{-1}\right)\right) \qquad (1.90)$$

as $|m| \to \infty$, and $m > 0$ corresponds to c_m for $s_\beta = -1$, while $m < 0$ corresponds to c_m for $s_\beta = 1$.

We introduce a function $g(\xi)$, associated in the sense of Borel [5,47] with the function $f^E(w_\pm)$:

$$g(\xi) = \sum_{n=0}^{\infty} n! \frac{c_n}{\xi^{n+1}}. \tag{1.91}$$

The function $g(\xi)$ is regular in the domain $|\xi| > \sigma$ [5,47], where σ is the degree of the entire function $f^E(w_\pm)$ (see Section 1.2.1).

We rewrite expansion (1.29) as

$$U^1(r, \varphi) = \sum_{m=0}^{\infty} a_m (-i)^m H_m^{(2)}(kr) e^{im\varphi} + \sum_{m=1}^{\infty} a_{-m} (-i)^m H_m^{(2)}(kr) e^{-im\varphi}$$
$$= F_1(r, \varphi) + F_2(r, \varphi). \tag{1.92}$$

It is known that the asymptotic relation

$$H_m^{(2)}(kr) = \frac{i}{\pi}(m-1)! \left(\frac{2}{kr}\right)^m \left(1 + O\left(\frac{1}{m}\right)\right)$$

holds as $m \to \infty$ [35,40].

Therefore, it is clear that the convergence of the series representing the functions $F_1(r, \varphi)$ and $F_2(r, \varphi)$ is equivalent to the convergence of the two series

$$\hat{F}_+(p_1) = \frac{i}{\pi} \sum_{m=1}^{\infty} a_m (m-1)! p_1^{-m}, \quad \hat{F}_-(p_2) = \frac{i}{\pi} \sum_{m=1}^{\infty} a_{-m} (m-1)! p_2^{-m}, \tag{1.93}$$

where $p_{1,2} = \frac{ikr}{2} e^{\mp i\varphi}$.

Comparing the power series (1.93) with the series (1.91), we see that the domains of regularity of the functions $\hat{F}_\pm$ and g coincide, i.e., the functions $\hat{F}_\pm$ are regular in the domain $|p_{1,2}| = \frac{kr}{2} > \sigma$. Therefore, we see that the expansion (1.29) holds in the domain

$$\frac{kr}{2} > \sigma. \tag{1.94}$$

Since the principal part of expansion (1.29) is a power series (see (1.93)), there is at least one singular point of the function $U^1(r, \varphi)$ on the circle $r = \frac{2\sigma}{k}$. However, as previously noted, the function $U^1(r, \varphi)$, which is the solution of the homogeneous Helmholtz equation in the domain $\mathbb{R}^2 \backslash \bar{D}$, is a real analytic function in this domain. Therefore, the singular points of the function can be located only in the interior of the domain D or on its boundary S. Relation (1.29) allows us to continue the solution

$U^1(r, \varphi)$ of the Helmholtz equation beyond the boundary of the initial domain of its definition, namely into the interior of the domain D to the circle $r = \frac{2\sigma}{k}$. As previously noted, the assumption that the series (1.29) and its three-dimensional analog converge to the scatterer boundary S is called the Rayleigh hypothesis [9,10,41,53,54]. It follows from the above that the Rayleigh hypothesis is satisfied only if the circle (sphere in the three-dimensional case) $r = \frac{2\sigma}{k}$ containing all singularities of the wave field lies inside the scatterer boundary S.

It follows from (1.54) and (1.90) that

$$a_{|m|} = \frac{\sigma_+^m}{|m|!}\left(1 + O\left(m^{-1}\right)\right). \qquad (1.95)$$

The estimate (1.95) allows us to prove an analog of the Liouville theorem for the wave fields, namely, if the wave field (i.e., the solution of the external boundary-value problem for the Helmholtz equation) is analytic everywhere, then it is equal to zero.

Indeed, if the wave field has no singularities, then $\sigma = 0$, namely, starting from some $|m| > N(\varepsilon)$, the coefficients of the series (1.29) satisfy the inequality $|a_m| < \varepsilon$, i.e., are arbitrarily small. However, since, by assumption, the field is regular everywhere including the origin, all coefficients a_m must be zero for $|m| \leq N(\varepsilon)$, and this means that the wave field is zero everywhere.

Similarly, we can show [37,41] that, in the three-dimensional case, the series (1.33) and (1.34) converge in the domain $\frac{kr}{2} > \sigma$, and in the three-dimensional case [41],

$$\sigma = \varlimsup_{R \to \infty} \frac{\ln \max_{|w|=R, \varphi}|f^E(w, \varphi)|}{R}, \qquad (1.96)$$

where (let us recall) $R = |w|$.

Now we consider representations (1.36) and (1.37) of the wave field $U^1(\vec{r})$ by integrals of plane waves and discuss representation (1.36) in more detail. The integral (1.36) converges or does not converge depending on the behavior of the integrand in a neighborhood of the points $-i\infty$, $\pi + i\infty$. Near these points, we have

$$Re\{-ikr\cos(\varphi - \psi)\} = ky\cos\alpha\,\text{sh}\beta = -\frac{ky}{2}\left(R - \frac{1}{R}\right).$$

Therefore, with (1.45) and (1.52) taken into account, we see that the integral (1.36) (with the upper signs) converges absolutely and uniformly for

$$y > \frac{2\sigma_S}{k}. \tag{1.97}$$

It follows from the definition of the upper limit that the integral (1.36) (with the upper signs) diverges for $y < 2\sigma_S/k$. On the line $y = 2\sigma_S/k$, the function $U^1(r, \varphi)$ has at least one singular point lying in the interior of the domain D or on its boundary S. If the coordinate axes are rotated by an angle γ in the positive direction, then using (1.52a), we can similarly show that the integral (1.36) with the upper signs converges absolutely and uniformly in the half-plane $y > \frac{2\sigma_S(0)}{k}$, and the integral with lower signs converges absolutely and uniformly in the half-plane $y < -\frac{2\sigma_S(\pi)}{k}$, while the integral (1.35) converges in the domain $\mathbb{R}^2 \backslash \bar{B}_0$, where $\bar{B}_0$ is the convex envelope of the singularities of the function $U^1(r, \varphi)$ continuation into the interior of the domain D.

It obviously follows from the above that $\bar{B}_0$ is the least closed convex set containing all singularities of the wave field.

Thus, the integral (1.35) allows us analytically to continue the function $U^1(r, \varphi)$, which is the solution of the homogeneous Helmholtz equation satisfying the prescribed boundary conditions on a closed contour S, into the domain D inside S to the boundary $\bar{B}_0$.

We can similarly show [41] that the integral (1.37) converges in the domain

$$z \equiv r\cos\theta > \frac{2\sigma_S}{k}, \tag{1.98}$$

where σ_S can now be determined, for example, from (1.64). Recall that this relation holds if $U^0(\vec{r})$ is the plane wave field and allows one to calculate σ_{S0}. But if the initial field is a point source, then

$$\sigma_S = \max\{\sigma_{S0}, 2z_0/k\},$$

where z_0 is the applicate of the source image.

By analogy with the two-dimensional case, considering the asymptotic behavior of the function $\hat{f}(\alpha, \beta; \theta, \varphi)$, we can show that the integral (1.38) converges in the domain $\mathbb{R}^3 \backslash \bar{B}_0$, where, as previously, $\bar{B}_0$ is the convex envelope of singularities of the wave field continuation.

Finally, we consider the wave field representation (1.40) by a series of plane waves. Completely repeating the reasoning about the integral (1.36), we can easily show [9,45] that the series (1.40) converges absolutely and uniformly for

$$y > \frac{2\sigma_S}{k}$$

1.2.5 Relationship Between the Asymptotics of the Pattern on the Complex Plane of its Argument and the Field Behavior near Singular Points

It is well known [55] that the asymptotics of the Fourier image determines the character of singularities of the Fourier transform. Similarly, the asymptotics of the scattering pattern, which is the spectral function in representations (1.35) and (1.36) on the complex plane of its argument, determines the character of singularities of the wave fields.

We illustrate the previously mentioned concepts and ideas by an example of the two-dimensional diffraction problem. To estimate the field character near the singular points, it is necessary to estimate the integral (1.46) as $R \equiv |w| \to \infty$ more precisely than in (1.48). Such an estimate can be obtained by the saddle-point method. The saddle points, i.e., the critical points of the phase function

$$F(\varphi) \equiv \frac{ik\rho(\varphi)}{2} \, e^{is_\beta(\varphi - \alpha)},$$

are determined by the condition

$$F'(\varphi)|_{\varphi = \varphi_0} \equiv \left[\frac{ik\rho'(\varphi_0)}{2} - \frac{s_\beta k \rho(\varphi_0)}{2} \right] e^{is_\beta(\varphi_0 - \alpha)} = 0,$$

i.e., by the equations

$$\frac{\rho'(\varphi_0)}{\rho(\varphi_0)} = -is_\beta, \quad e^{is_\beta \varphi_0} = 0 \tag{1.99}$$

(cf. (1.56)).

The standard estimate implies

$$f^E(w_\pm) \sim R^{\pm 1/2} \sum_m c(\varphi_{0m}) \exp[RF(\varphi_{0m})], \tag{1.100}$$

where the sum is taken over all saddle points, $c(\varphi_{0m})$ are constants, the "minus" sign in the index corresponds to the Dirichlet boundary condition on the scatterer boundary, and

$$F(\varphi_0) = \frac{ik\rho(\varphi_0)}{2} e^{is_\beta(\varphi_0 - \alpha)}. \tag{1.101}$$

It follows from (1.51), (1.52) and (1.100), (1.101) that

$$\sigma = \max_{\varphi_0, \alpha, s_\beta} Re\{F(\varphi_0)\}, \tag{1.102}$$

$$\sigma_S = \max_{\varphi_0, w = \pm R} Re\{F(\varphi_0)\}. \tag{1.103}$$

In (1.102) and (1.103), the maximum is taken over the roots of equations (1.99) that are associated with singular points inside the contours $C^{\pm}$ on the complex plane that correspond to the boundary S under the mappings $\zeta^{\pm} = \rho(\varphi)e^{\pm i\varphi}$ (see Section 1.2.2).

For example, let us consider the solution of the problem of plane wave diffraction at an elliptic cylinder with the Dirichlet boundary condition. First, we determine the domain of regularity of the scattered field. We previously noted that the integral (1.35) converges absolutely and uniformly for $y > \frac{2\sigma_S(\gamma)}{k}$, and the integral satisfies the homogeneous Helmholtz equation and the condition of radiation at infinity, i.e., it is a regular function in this domain.

It was shown in Section 1.2.3.1 that, in the case under study, the roots of the first equation in (1.99) are determined by the relations

$$\exp\left(is_\beta\varphi_{0m}\right) = \pm\frac{\varepsilon}{\sqrt{2-\varepsilon^2}}e^{im\pi}, \quad m = 0, 1,$$

so that

$$F(\varphi_{0m}) = \frac{ikb\varepsilon}{2}e^{-is_\beta\alpha + im\pi}, \quad m = 0, 1, \tag{1.104}$$

which implies (see above) that $\sigma_S(0) = \sigma_S(\pi) = 0$ and $\sigma_S(\pm\frac{\pi}{2}) = kb\varepsilon/2$. Therefore, the field generated by the plane wave scattering at an ellipse is regular everywhere outside the interfocus segment $-b\varepsilon \le x \le b\varepsilon$.

According to (1.100) and (1.104) [9, Chapter 2], [53,54], we have

$$f^E(w) \sim R^{-1/2}\sum_{m=0}^{1} c(\varphi_{0m})\exp\left[ikb\varepsilon\frac{w}{2}e^{im\pi}\right]. \tag{1.105}$$

It follows from (1.105) that the integral (1.36) converges absolutely for $y \ge 0$. If we take the opposite sign of ψ in (1.36) (which corresponds to the rotation of the coordinate axes by the angle $\gamma = \pi$), then the integral converges in the half-plane $y \le 0$. Because $f(\psi) \ne f(-\psi)$ in the general case, we have $U^1(x, y+0) \ne U^1(x, y-0)$ (we preserved the same letter for the diffraction field in the Cartesian coordinates) on the interval $-b\varepsilon \le x \le b\varepsilon$, i.e., the diffraction field experiences a jump when passing through the interfocus segment. Apropos, in the case of symmetric irradiation (where the incident wave propagates along the major axis of the ellipse), we have $f(\psi) = f(-\psi)$, and hence the diffraction field is everywhere continuous. The normal derivative $\partial U^1/\partial n$ also experiences a jump when passing through the interfocus segment. Let us determine the character of the diffraction field singularity near the ellipse foci [53,54]. For this, we rotate the coordinate axes by $90°$. Expression (1.105) is then transformed as follows (α must be replaced by $\alpha + \pi/2$):

$$f^E(w_\pm) \sim R^{-1/2} \sum_{m=0}^{1} c(\varphi_{0m}) \exp\left[-kbe\frac{w}{2}e^{im\pi}\right]. \qquad (1.105a)$$

We rewrite the integral (1.36) in the Cartesian coordinates $x = r\cos\varphi$, $y = r\sin\varphi$ by changing the variable by the formula $v = k\cos\psi$. Then for $y > 2\sigma_S/k$, we have

$$U^1(x, y) = \frac{1}{\pi}\int_{-\infty}^{\infty} \widetilde{f}(v) \frac{\exp\left(-ivx - iy\sqrt{k^2 - v^2}\right)}{\sqrt{k^2 - v^2}} dv, \qquad (1.106)$$

where $\widetilde{f}(v) = f(\psi(v))$. On the line $x = 0$; $\partial/\partial n = \partial/\partial x$. Therefore, (1.106) implies

$$\left.\frac{\partial U^1}{\partial x}\right|_{x=0} = \frac{1}{\pi i}\int_{-\infty}^{\infty} \frac{\widetilde{f}(v)v}{\sqrt{k^2 - v^2}} e^{-iy\sqrt{k^2 - v^2}} dv. \qquad (1.107)$$

As $|v| \to \infty$, $k\cos\psi \approx kw/2$, i.e., $w \approx 2v/k$. Thus, with (1.105a) taken into account, for the infinite remainders in the integral (1.107) with $y = b\varepsilon + t$, $t > 0$, we obtain

$$\text{const}\int_p^{\infty} \frac{e^{-vt}}{v^{1/2}} dv = \text{const}\frac{\Gamma(1/2)}{t^{1/2}} - G(t), \qquad (1.108)$$

where $p > 0$ is a sufficiently large number and

$$G(t) = \text{const}\int_0^p \frac{e^{-vt}}{v^{1/2}} dv$$

is a bounded function. In (1.108), we used the relation [35]

$$t^{-\xi}\Gamma(\xi) = \int_0^{\infty} v^{\xi-1} e^{-vt} dv. \qquad (1.109)$$

It follows from (1.108) that near the ellipse foci, i.e., for $x = 0$, $y = \pm b\varepsilon + t$, the field has a singularity of the form $t^{-1/2}$, i.e., the same singularity as the electric current singularity near the edge of an ideally conducting band.

If the degeneration[1] of the phase function $F(\varphi)$ at stationary points is of the qth order, i.e., if

---

[1] To simplify our consideration, we assume that the degeneration order is the same at all stationary points φ_{0m}, which is obviously unnecessary.

$$F'(\varphi_{0m}) = F''(\varphi_{0m}) = \cdots = F^{q-1}(\varphi_{0m}) = 0, \quad F^q(\varphi_{0m}) \neq 0,$$

then the character of singularities at singular points is different [9, Chapter 2]. In this case, for the function $f^E(w)$, instead of (1.105), we have

$$f^E(w) \sim \frac{2\Gamma(1/q)(q!)^{1/q}}{qR^{1/q}} \sum_m c(\varphi_{0m}) \frac{e^{RF(\varphi_{0m}) + i\kappa}}{|F^{(q)}(\varphi_{0m})|^{1/q}}, \tag{1.110}$$

where $\kappa = -(1/q)\arg F^{(q)}(\varphi_{0m}) - \pi/q$.

An example of a figure whose phase function of the pattern has such properties is a q-foil (see above).

We substitute the asymptotic expansion (1.110) of the function $f^E(w)$ in the integral

$$\int_0^{\infty e^{-i\gamma}} f^E(w) e^{-w\xi} dw \equiv G(\xi). \tag{1.111}$$

The character of singular points of the function $G(\xi)$ for $\xi = \frac{ikr}{2}e^{i\varphi}$ and of the wave field $U^1(r,\varphi)$ is the same [9, Chapter 2]. After substitution, we obtain

$$G^{pr}(\xi) = \sum_m c_m \int_0^{\infty e^{-i\gamma}} w^{-1/q} e^{-w(\xi - \zeta_{0m})} dw, \quad w = Re^{-i\gamma},$$

where c_m are constants, $\zeta_{0m} = \frac{ik\rho(\varphi_{0m})}{2}e^{i\varphi_{0m}}$, φ_{0m} are the saddle points of the phase function $F(\varphi)$, and $G^{pr}(\xi)$ is the part of the function $G(\xi)$ containing its principal singularities. We use (1.109) to obtain

$$G^{pr}(\xi) = \Gamma\left(1 - \frac{1}{q}\right) \sum_m \frac{c_m}{(\xi - \zeta_{0m})^{1-1/q}}$$

or, returning to the variables (r, φ),

$$G^{pr}\left(\frac{ikr}{2}e^{i\varphi}\right) = \frac{\Gamma(1 - 1/q)}{\left(\frac{ir}{2}\right)^{1-1/q}} \sum_m \frac{c_m}{(re^{i\varphi} - r_{0m}e^{i\varphi_{0m}})^{1-1/q}},$$

where (r_{0m}, φ_{0m}) are the coordinates of singular points.

In this chapter, we considered the basic analytic representations of wave (diffraction) fields, determined the exact boundaries of the existence domains of these representations, and showed that these boundaries are strictly conditioned by the geometry of the set of the analytic continuation singularities of the diffraction field beyond the domain, where we seek the solution of the boundary-value diffraction problem and which can be established *a priori* (before solving this problem) [9,10,30,34,37,41,42,45,48,50,53,54]. In the subsequent chapters, we discuss the methods for solving the diffraction problems, which are essentially based on this *a priori* information.

The second chapter deals with the methods of auxiliary currents and sources [16,26,30,52,54,56–77]. These methods are based on representing the wave field by (auxiliary) sources lying in the nonphysical region of the space, i.e., inside the scatterer for the scattered field and in the domain external with respect to the scatterer for the internal field (if the problem is solved with conjugation conditions on the boundary). One of the key points in the discussed methods is the choice of the support of the auxiliary sources. A correct choice of the support based on the use of *a priori* information about the singularities of the wave field continuation into the nonphysical region not only guarantees that the corresponding integral equations have solutions but also permits constructing the most optimal numerical algorithms.

In the third chapter, we discuss the NFMs and the T-matrix methods [22,23,78–82]. These methods are based on the classical relation of the theory of potential, which is called the null zero field relation in the literature concerned with the theory of diffraction [5,19]. Under the assumption that this relation is satisfied on some (auxiliary) surface inside the scatterer (when solving the problem with conjugation conditions on the interface both inside and outside the scatter), the boundary-value problem is reduced to the Fredholm integral equation of the first type with respect to the distribution of sources on the scatterer boundary. The NFM version, called the T-matrix method, where the auxiliary surface is a sphere, is most popular [19,20]. Until recently, it was assumed that there are no restrictions on the choice of the auxiliary surface except the requirement that it be not at resonance [5]. However, this was shown to be untrue [22,23,78–82]. Because some analytic representations of the diffraction field are used in any numerical realization of the method, the auxiliary surface must be chosen taking into account the geometry of the set of the analytic continuation singularities of the wave field into the nonphysical region. Moreover, it was shown that the most stable and fast-operating algorithms are realized when the auxiliary surface is constructed by using the analytic deformation of the scatterer boundary to the wave field singularities, just as in the method of auxiliary currents.

The fourth chapter deals with the method of CBCM [13,14,21,73,84–88]. In this method, it is essential to use the fact that the diffraction field, which is a solution of the homogeneous Helmholtz equation (homogeneous system of Maxwell equations), is a real analytic function [5,6]. Thus, if a boundary condition is satisfied on the scatterer boundary S, then this condition is also approximately satisfied on a surface S_δ sufficiently close to the surface S. This allows one, writing the boundary condition on the surface S_δ, to obtain the Fredholm integral equation of the first type with a smooth kernel for determining the sources on the scatterer surface S. Such equations are preferable to the traditional singular integral equations, because the algorithms for solving them are significantly simpler and "faster." The analyticity of the wave field makes such a problem quite well posed [73]. The point is that the existence and uniqueness theorems can be proved for the solution of the corresponding integral equation under certain restrictions on the choice of the surface S_δ [13]. The third requirement that the statement of the problem be well posed, i.e., the requirement of stability of its solution [4,89] is not satisfied here. But this fact is not of principal importance, because the source distribution on the scatterer surface is a quantity that cannot be observed physically, and hence this distribution is not interesting on its own but the calculations of all functionals of this distribution, including the diffraction field in any neighborhood of the scatterer boundary, are quite stable [73]. Moreover, in the CBCM framework, the initial boundary-value problem can be reduced to solving the Fredholm integral equation of the second type [87,88], which removes all problems of correctness at all.

The fifth chapter deals with the pattern equations method [42,44,51,94–131]. In the framework of this method, the initial boundary-value diffraction problem reduces to solving an integral–operator equation (or a system of such equations) with respect to the wave field pattern. *A priori* analytic information is used to justify the method and to determine the limit of its applicability.

CHAPTER 2

Methods of Auxiliary Currents and Method of Discrete Sources

The method of auxiliary currents (MAC), as well as the method of discrete sources (MDS), which is one of its realizations, is one of most popular methods for solving the boundary-value problems of diffraction theory (e.g., see Refs. [25, 27, 28]). In this chapter, we briefly describe both of these methods, their justification, and an effective modification of these methods.

2.1 EXISTENCE AND UNIQUENESS THEOREMS

As previously noted, the MAC is based on representing the wave field in terms of wave potentials of the form (1.2a).

Assume that it is first required to solve the homogeneous Dirichlet boundary-value problem

$$\left(U^0 + U^1\right)\big|_S = 0, \tag{2.1}$$

where S is the scatterer boundary, U^0 is the incident field, and U^1 is the secondary (scattered) field satisfying the radiation condition at infinity. The use of representation (1.2a) reduces this problem to solving the integral equation of the first type with a smooth kernel [16]

$$\int_{\Sigma} I\left(\vec{r}_\Sigma\right) G_0\left(\vec{r}_S; \vec{r}_\Sigma\right) d\sigma = U^0\left(\vec{r}_S\right), \quad \vec{r}_S \in S, \tag{2.2}$$

where Σ is an auxiliary closed surface in the domain D.

It follows from the above that Equation (2.2) is solvable only under the condition that Σ surrounds all singularities of the analytic continuation of the function $U^1\left(\vec{r}\right)$ into a domain inside S. More precisely, the following existence theorem (Theorem 1) holds [10, 16, 49].

Mathematical Modeling in Diffraction Theory
http://dx.doi.org/10.1016/B978-0-12-803728-7.00002-6

Theorem 1 *Assume that a simple closed surface Σ satisfies the condition that k is not an eigenvalue of the internal homogeneous Dirichlet problem for a domain inside Σ. Then Equation (2.2) is solvable if and only if Σ surrounds all singularities of the solution $U^1\left(\vec{r}\right)$ of boundary-value problem (2.1) for the Helmholtz equation.*

Indeed, if Equation (2.2) is solvable, i.e., if there is an integrable function $I\left(\vec{r}_\Sigma\right)$ such that (2.2) is satisfied, then it follows from the properties of the function $G_0\left(\vec{r}; \vec{r}_\Sigma\right)$ that the integral in the left-hand side of (2.2) with the function $G_0\left(\vec{r}_S; \vec{r}_\Sigma\right)$ replaced by $G_0\left(\vec{r}; \vec{r}_\Sigma\right)$, where $\vec{r}$ is the radius vector of a point outside Σ, satisfies the homogeneous Helmholtz equation everywhere outside Σ and, therefore, has no singularities.

Conversely, assume that Σ surrounds all singularities of the wave field inside $\bar{D}$ and hence the field $U^1\left(\vec{r}\right)$ can be continued into the interior of D to Σ. Let $U^i\left(\vec{r}\right)$ be a solution of the interior (for a domain inside Σ) Dirichlet problem with the value $U^1\left(\vec{r}\right)|_\Sigma$ given on Σ. Then the function $\widetilde{U}\left(\vec{r}\right)$, which is equal to $U^1\left(\vec{r}\right)$ outside Σ and to $U^i\left(\vec{r}\right)$ inside Σ, is continuous on the entire space and its normal (to Σ) derivative experiences a jump when passing through Σ. The value of this jump can be taken as $I\left(\vec{r}_\Sigma\right)$.

The following uniqueness theorem (Theorem 2) holds.

Theorem 2 *Under the conditions of Theorem 1, Equation (2.2) has only a unique solution.*

Assume that Equation (2.2) has two solutions: $I_1\left(\vec{r}_\Sigma\right)$ and $I_2\left(\vec{r}_\Sigma\right)$. Then their difference is a solution of the corresponding homogeneous equation

$$\int_\Sigma \mu\left(\vec{r}_\Sigma\right) G_0\left(\vec{r}_S; \vec{r}_\Sigma\right) d\sigma = 0, \tag{2.3}$$

where $\mu\left(\vec{r}_\Sigma\right) = I_2\left(\vec{r}_\Sigma\right) - I_1\left(\vec{r}_\Sigma\right)$.

Let $U\left(\vec{r}\right) = \int_\Sigma \mu\left(\vec{r}_\Sigma\right) G_0\left(\vec{r}, \vec{r}_\Sigma\right) ds$. Then it follows from the properties of the function G_0 that $U\left(\vec{r}\right)$ is a solution of the homogeneous Helmholtz equation everywhere outside Σ:

$$\Delta U\left(\vec{r}\right) + k^2 U\left(\vec{r}\right) = 0. \tag{2.4}$$

We consider a domain V_e bounded from the outside by the sphere S_R and from the inside by the scatterer surface S and let the radius R of the sphere S_R tend to infinity (see Fig. 2.1).

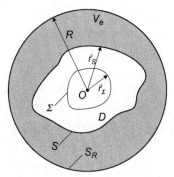

Figure 2.1 To the proof of Theorem 2.

We multiply Equation (2.4) by $\bar{U}$ and integrate it over V_e to obtain

$$\int_{V_e} \left(\bar{U}\Delta U + k^2 |U|^2 \right) dv = 0, \tag{2.5}$$

or

$$k^2 \int_{V_e} |U|^2 dv = -\int_{V_e} [\mathrm{div}(\bar{U}\nabla U) - \nabla \bar{U}\nabla U] dv. \tag{2.6}$$

We use the Ostrogradsky theorem to replace the integration of the divergence over V_e by the integration over the boundary of V_e, i.e., over $\{S + S_R\}$. We obtain

$$k^2 \int_{V_e} |U|^2 dv = \int_{V_e} \nabla \bar{U}\nabla U dv - \int_{S+S_R} \bar{U}\frac{\partial U}{\partial n} ds. \tag{2.7}$$

Note that that equality $U|_S = 0$ implies $\int_S \bar{U}\frac{\partial U}{\partial n} ds = 0$, and because of the condition at infinity, we have $\lim_{R\to\infty} \int_{S_R} \bar{U}\frac{\partial U}{\partial n} ds_R = 0$ for the function $U(\vec{r})$. Therefore, the last integral in (2.7) tends to zero as $R \to \infty$ and (2.7) then becomes

$$k^2 \int_{V_e} |U|^2 dv = \int_{V_e} \nabla \bar{U}\nabla U dv. \tag{2.8}$$

Let k be a complex number (i.e., $k^2 = \omega^2 \varepsilon \mu$, $\varepsilon = \varepsilon_1 - i\varepsilon_2$), which corresponds to the fact that there are losses of conductivity in the environment. We take the imaginary part of (2.8):

$$-\omega^2 \mu \varepsilon_2 \int_{V_e} |U|^2 dv = 0. \tag{2.9}$$

It follows from (2.9) that $U \equiv 0$ in the domain V_e.

The function U is analytic everywhere outside Σ (all singularities of the solution of the boundary-value problem (2.1) lie inside the surface Σ), and hence, $U(\vec{r}) \equiv 0$ everywhere outside Σ, which also implies (by continuity) that $\frac{\partial U}{\partial n}\big|_\Sigma = 0$, where $\vec{n}$ is the outer normal on Σ.

We consider the internal homogeneous Dirichlet problem for the domain inside the surface Σ:

$$\begin{cases} \Delta U^i(\vec{r}) + k^2 U^i(\vec{r}) = 0, \\ U^i(\vec{r})\big|_\Sigma = 0. \end{cases} \tag{2.10}$$

For each surface Σ, there is a discrete set of values of k for which this problem has a nontrivial solution. However, a small variation in the parameter k can easily ensure that the surface Σ is not at resonance. Thus, problem (2.10) has only a trivial solution. Therefore, $\partial U^i/\partial n = 0$. Then we have

$$I \equiv \left(\frac{\partial U}{\partial n} - \frac{\partial U^i}{\partial n} \right) \bigg|_\Sigma = 0.$$

Thus, $I_1 = I_2$, which means that Equation (2.2) can have only one solution, as required.

2.2 SOLUTION OF THE MAC INTEGRAL EQUATION AND THE MDS

We consider the technique for solving equations of the form (2.2). For simplicity, we consider only the two-dimensional case. In what follows, it is convenient to parameterize the contours S and Σ. We assume that both contours are star shaped. Then the polar angle can be regarded as a parameter. Thus, we set

$$r_S = r(\alpha), \quad 0 \le \alpha < 2\pi; \quad r_\Sigma = \rho(\theta), \quad 0 \le \theta < 2\pi,$$
$$\frac{i}{4} I(\vec{r}_\Sigma) d\sigma = \mu(\theta) d\theta; \quad -U^0(\vec{r}_S) = \psi(\alpha).$$

As a result, Equation (2.2) becomes

$$\int_0^{2\pi} \mu(\theta) H_0^{(2)}\left(k|\vec{r}_S - \vec{r}_\Sigma|\right) d\theta = \psi(\alpha), \quad \alpha \in [0, 2\pi[, \qquad (2.11)$$

where we now have

$$|\vec{r}_S - \vec{r}_\Sigma| = \left[r^2(\alpha) + \rho^2(\theta) - 2r(\alpha)\rho(\theta)\cos(\alpha - \theta)\right]^{1/2}. \qquad (2.12)$$

The most popular method for solving such equations is the method of moments where the unknown function $\mu(\theta)$ is sought as the expansion

$$\mu(\theta) = \sum_n c_n f_n(\theta), \qquad (2.13)$$

with respect to a complete system of functions. To obtain the exact representation of $\mu(\theta)$, we must take infinitely many terms in (2.13). However, in practice, when a finite accuracy of the approximation is prescribed, one usually takes a finite number N of terms in the right-hand side of (2.13). We represent the function $\mu(\theta)$ in Equation (2.11) in this way and obtain

$$\sum_{n=1}^N c_n L f_n \approx \psi,$$

where L denotes the integral operator in (2.11). Further, we project this equality on a certain basis $(\xi_m | m = 1, \ldots, N)$ and obtain the following system of linear algebraic equations (SLAE) for the unknown coefficients c_n:

$$\sum_{n=1}^N c_n \langle \xi_m, L f_n \rangle = \langle \xi_m, \psi \rangle, \quad m = 1, \ldots, N. \qquad (2.14)$$

Solving this system, we obtain the coefficients c_n and hence solve the integral equation (2.11). The delta function $\xi_m = \delta(\alpha - \alpha_m)$, where α_m are discretization points of the interval $\alpha \in [0, 2\pi[$ (collocation points), is often taken as ξ_m. Therefore, the method for solving equations of the form (2.11) is called the collocation method [43].

The most simple method for approximating the unknown function in equations of the form (2.11) is to represent it as an expansion with respect to a piecewise constant basis:

$$f_n(\theta) = \begin{cases} 1, & \theta \in [2(n-1)\pi/N, 2n\pi/N), \\ 0, & \theta \notin [2(n-1)\pi/N, 2n\pi/N). \end{cases}$$

Then SLAE (2.14) becomes

$$\sum_{n=1}^{N} c_n \int_{2(n-1)\pi/N}^{2n\pi/N} K(\alpha_m, \theta) d\theta = \psi(\alpha_m), \quad m = 1, \ldots, N, \tag{2.15}$$

where $\alpha_m = \dfrac{2\pi}{N}(m - 1/2)$ and

$$K(\alpha_m, \theta) = H_0^{(2)} \left(k\sqrt{r^2(\alpha_m) + \rho^2(\theta) - 2r(\alpha_m)\rho(\theta)\cos(\alpha_m - \theta)} \right).$$

If the integrals in SLAE (2.15) are replaced by the product of the integrand value at the middle of the interval by the interval length, then we obtain the algebraic system of the MDS (see below)

$$\frac{2\pi}{N} \sum_{n=1}^{N} c_n K(\alpha_m, \theta_n) = \psi(\alpha_m), \tag{2.16}$$

where

$$\theta_n = \frac{2\pi}{N}\left(n - \frac{1}{2}\right).$$

If the unknown function is approximated by linear splines, then we have

$$\mu(\theta) = \frac{N}{2\pi} \sum_{n=0}^{N} c_n \cdot \phi_{1n}(\theta),$$

where

$$\phi_{1n}(\theta) = \begin{cases} \theta - \beta_n, & \theta \in [(n-1)2\pi/N, 2n\pi/N), \\ \beta_{n+1} - \theta, & \theta \in [2n\pi/N, 2(n+1)\pi/N), \\ 0, & \text{else} \end{cases} \quad \beta_n = \frac{2(n-1)\pi}{N}, \quad n = \overline{1, N-1},$$

$$\phi_{10}(\theta) = \begin{cases} \beta_0 - \theta, & \theta \in [0, 2\pi/N), \\ 0 & \text{else} \end{cases} \quad \beta_0 = \frac{2\pi}{N},$$

$$\phi_{1N}(\theta) = \begin{cases} \theta - \beta_N, & \theta \in [2(N-1)\pi/N, 2\pi), \\ 0, & \text{else} \end{cases} \quad \beta_N = \frac{2(N-1)\pi}{N}.$$

As a result, the SLAE becomes

$$\frac{N}{2\pi}\left\{ c_0 \int_0^{2\pi/N} K(\alpha_m,\theta)(\beta_0-\theta)\mathrm{d}\theta \right.$$

$$+ \sum_{n=1}^{N-1} c_n \left(\int_{(n-1)2\pi/N}^{2n\pi/N} K(\alpha_m,\theta)(\theta-\beta_n)\mathrm{d}\theta + \int_{2n\pi/N}^{(n+1)2\pi/N} K(\alpha_m,\theta)(\beta_n-\theta)\mathrm{d}\theta \right)$$

$$\left. + c_N \int_{(N-1)2\pi/N}^{2\pi} K(\alpha_m,\theta)(\theta-\beta_N)\mathrm{d}\theta \right\} = \psi(\alpha_m), \quad \alpha_m = \frac{2m\pi}{N}, m = \overline{0,N}.$$

$$(2.17)$$

In [56], the splines were studied with the purpose of solving integral equations of the form (2.11). In particular, it was shown that the use of linear splines allows one to increase (almost by an order of magnitude) the accuracy of calculations compared with the piecewise constant approximation of the unknown function and by more than two orders of magnitude compared with the MDS.

Nevertheless, the MDS is widely used to solve the wave diffraction problems (e.g., see Refs. [25, 27]). This can mainly be explained by the algorithmic simplicity of the method. For our further purposes, it is more convenient to rewrite the algebraic system (2.16) as

$$\sum_{n=1}^{N} c_n H_0^{(2)}\left(k|\vec{r}_{S_m} - \vec{r}_{\Sigma_n}|\right) = -U^0\left(\vec{r}_{S_m}\right), \quad m = 1,\dots,N, \qquad (2.16a)$$

where $\vec{r}_{\Sigma_n}$ is the radius vector of a point on the contour Σ corresponding to the number n and $\vec{r}_{S_m}$ are radius vectors of collocation points on S.

The MDS was proposed by Kupradze and Aleksidze [17, 18]. In particular, they proved the following theorem.

Theorem 3 *If Σ is a nonresonance contour, the points Ω_n with radius vectors $\vec{r}_{\Sigma_n}$ are located everywhere dense on Σ, i.e., $\overline{\{\Omega_n\}|_{n=0}^{\infty}} = \Sigma$, and $\vec{r}$ is the radius vector of an arbitrary point on S, then the system of functions $\left\{ H_0^{(2)}\left(k|\vec{r} - \vec{r}_{\Sigma_n}|\right)\right\}_{n=0}^{\infty}$ is complete in $L^2(S)$.*

If in the left-hand side of (2.16a), $\vec{r}_{S_m}$ is replaced by the radius vector $\vec{r}$ of an arbitrary point outside Σ, then the left-hand side is an approximation of the diffraction field

$$U^1(\vec{r}) \cong \sum_{n=1}^{N} c_n H_0^{(2)}(k|\vec{r} - \vec{r}_{\Sigma_n}|). \tag{2.18}$$

Relation (2.18) automatically satisfies the Helmholtz equation and the radiation condition, and therefore, the accuracy of the obtained approximate solution can be estimated by the accuracy of the boundary condition satisfaction (the discrepancy) at the points of the contour S different from the collocation points. This method is algorithmically very simple and extremely fast operating.

The support Σ of auxiliary sources in (2.11) and (2.16) must be chosen according to Theorem 1. However, the choice of Σ in this theorem is rather ambiguous (there are infinitely many contours between S and the set of singularities), which can cause (and often rather serious) difficulties in the practical use of MDS. In the so-called modified method of discrete sources (MMDS) [26], the support of discrete sources is constructed by an analytic deformation of the scatterer boundary S.

2.3 RIGOROUS SOLUTION OF THE DIFFRACTION PROBLEM BY MAC [9, 16]

We consider the problem of diffraction of a plane electromagnetic wave at an ideally conducting elliptic cylinder in the case where the incident field electric vector is parallel to the cylinder axis. We assume that the major axis of the ellipse S, which is the cylinder cross-section, is oriented along the axis Ox. As shown in Chapter 1, the diffraction field in this case is analytic everywhere outside the interfocus segment $-b\varepsilon \le x \le b\varepsilon$. Thus, the diffraction field $U^1(\vec{r})$ can be represented by the integral

$$U^1(\vec{r}) = -\frac{1}{4i} \int_{\Sigma} \mu(\vec{r}_{\Sigma}) H_0^{(2)}(k|\vec{r} - \vec{r}_{\Sigma}|) \, d\sigma, \tag{2.19}$$

where Σ is an arbitrary (but not at resonance) closed Lyapunov contour surrounding the segment $-b\varepsilon \le x \le b\varepsilon$. For Σ we take an ellipse confocal with the initial ellipse S. In elliptic coordinates (ξ, η), we have

$$d\sigma = f\sqrt{ch^2\xi_0 - \cos^2\eta} \, d\eta,$$

where $\xi = \xi_0$ is the equation of the contour Σ and $f = b\varepsilon$. After the change

$$\frac{i}{4}\mu(\vec{r}_{\Sigma}) = I(\eta)\left(f\sqrt{ch^2\xi_0 - \cos^2\eta}\right)^{-1},$$

representation (2.19) becomes

$$U^1(\vec{r}) = \int_0^{2\pi} I(\eta') H_0^{(2)}(k|\vec{r} - \vec{r}'|)\,d\eta'. \tag{2.20}$$

Thus, the function $I(\eta')$ can be determined from the equation

$$\int_0^{2\pi} I(\eta') H_0^{(2)}(k|\vec{r} - \vec{r}'|)\,d\eta' = -e^{-i\vec{k}\vec{r}}, \quad M(\vec{r}) \in S, \tag{2.21}$$

where $\vec{k}\vec{r} = kr\cos(\varphi - \varphi_0)$ and (r, φ) are the observation point coordinates and φ_0 is the angle of the plane wave incidence calculated from the axis Ox.

In elliptic coordinates (here all notation symbols coincide, for example, with those in [40]),

$$e^{-i\vec{k}\vec{r}} = 2\sum_{n=0}^{\infty}[c_n Ce_n(\xi, q)ce_n(\eta, q)ce_n(\varphi_0, q) + d_n Se_n(\xi, q)se_n(\eta, q)se_n(\varphi_0, q)],$$
$$\tag{2.22}$$

$$H_0^{(2)}(k|\vec{r} - \vec{r}'|) = 2\sum_{m=0}^{\infty}(-1)^m\Big[c_m^2 Me_m^{(2)}(\xi, q)Ce_m(\xi_0, q)ce_m(\eta, q)ce_m(\eta', q)$$
$$+ d_m^2 Ne_m^{(2)}(\xi, q)Se_m(\xi_0, q)se_m(\eta, q)se_m(\eta', q)\Big], \quad \xi \geq \xi_0$$
$$\tag{2.23}$$

where $q = (kf/2)^2$, ce_n, se_n are even and odd angular Mathieu functions, and $Ce_n(\xi, q)$, $Se_n(\xi, q)$, $Me_n^{(2)}(\xi, q)$, $Ne_n^{(2)}(\xi, q)$ are modified Mathieu functions [40]. Substituting (2.22) and (2.23) into (2.21) and using the fact that the Mathieu functions ce_n and se_n are orthogonal on the interval $[0, 2\pi]$, we obtain

$$I(\eta') = -\frac{1}{\pi}\sum_{n=0}^{\infty}(-1)^n\left[\frac{ce_n(\eta', q)}{c_n}\frac{Ce_n(\xi_1, q)}{Me_n^{(2)}(\xi_1, q)}\frac{ce_n(\varphi_0, q)}{Ce_n(\xi_0, q)}\right.$$
$$\left. + \frac{se_n(\eta', q)}{d_n}\frac{Se_n(\xi_1, q)}{Ne_n^{(2)}(\xi_1, q)}\frac{se_n(\varphi_0, q)}{Se_n(\xi_0, q)}\right]$$

where $\xi = \xi_1$ is the equation of the cylinder cross-section contour S at which the waves diffract. If we substitute this relation into (2.20), then we obtain the following well-known expression for the diffraction field [40]:

$$U^1(\vec{r}) = -2\sum_{n=0}^{\infty} \left[c_n \frac{Ce_n(\xi_1, q)}{Me_n^{(2)}(\xi_1, q)} Me_n^{(2)}(\xi, q) ce_n(\varphi_0, q) ce_n(\eta, q) \right.$$

$$\left. + d_n \frac{Se_n(\xi_1, q)}{Ne_n^{(2)}(\xi_1, q)} Ne_n^{(2)}(\xi, q) se_n(\varphi_0, q) se_n(\eta, q) \right]. \tag{2.24}$$

If the contour Σ is contracted to the interval $[-f, f]$ of the axis Ox, then representation (2.19) of the diffraction field in the form of the simple layer potential is already incomplete and it is necessary to use representation (1.2a). Indeed, we consider the integral (1.36), where the upper signs are taken for $y > \frac{2\sigma_S(0)}{k}$, and the lower signs, for $y < -\frac{2\sigma_S(\pi)}{k}$. In the case under study, we have $\sigma_S(0) = \sigma_S(\pi) = 0$ (see Chapter 1) and the integrals (1.36) converge absolutely and uniformly for $y > 0$ and $y < 0$, respectively. Because $f(+\psi) \neq f(-\psi)$ in general, we have $U^1(x, y+0) \neq U^1(x, y-0)$, i.e., the diffraction field experiences a jump when passing through the axis Ox on the interval $[-f, f]$. The normal derivative $\partial U^1/\partial n$ also experiences a jump, and hence we must use representation (1.2a).

We write (1.2a) in the form that is more convenient for our further studies. To this end, we introduce the notation

$$\tilde{I}^e(\eta) \equiv f\sqrt{1 - \cos^2\eta}\, I^e(\eta) \equiv \frac{i}{4} \left[\frac{\partial U^1(\xi, \eta)}{\partial \xi} + \frac{\partial U^1(\xi, -\eta)}{\partial \xi} \right]_{\xi=0},$$

$$I^m(\eta) \equiv -\frac{i}{4} \left[U^1(\xi, \eta) - U^1(\xi, -\eta) \right]_{\xi=0}.$$

Now using the fact that

$$\frac{\partial}{\partial n} = \frac{1}{f\sqrt{ch^2\xi - \cos^2\eta}} \frac{\partial}{\partial \xi}, \quad d\sigma = f\sqrt{ch^2\xi - \cos^2\eta}\, d\eta,$$

we obtain from (1.2a):

$$U^1(\vec{r}) = \int_0^\pi \left[I^m(\eta') \frac{\partial}{\partial \xi'} H_0^{(2)}(k|\vec{r} - \vec{r}'|) + \tilde{I}^e(\eta') H_0^{(2)}(k|\vec{r} - \vec{r}'|) \right] d\eta'.$$

$$\tag{2.25}$$

Further, using the boundary condition $U^0 + U^1 = 0$ on S, i.e., for $\xi = \xi_1$, we must determine the quantities I^m and $\tilde{I}^e$. Because there are two unknown functions, we need two equations to determine them. We obtain the required equations if we note that the term with $\tilde{I}^e$ in (2.25) generates a field symmetric with respect to the line $y = 0$ and the term with I^m generates an antisymmetric field. Therefore, we divide the field $U^0(\vec{r})$ into the symmetric and antisymmetric parts:

$$U^0(\vec{r}) = e^{-ikr\cos(\varphi - \varphi_0)} = e^{-ikx\cos\varphi_0}\cos(ky\sin\varphi_0) - ie^{-ikx\cos\varphi_0}\sin(ky\sin\varphi_0).$$

Now we can obtain the unknown quantities I^m and $\tilde{I}^e$ from the two integral equations

$$\int_0^\pi I^m(\eta')\frac{\partial}{\partial\xi'}H_0^{(2)}\left(k|\vec{r} - \vec{r}'|\right)d\eta' = ie^{-ikx\cos\varphi_0}\sin(ky\sin\varphi_0), \qquad (2.26a)$$

$$\int_0^\pi I^e(\eta')H_0^{(2)}\left(k|\vec{r} - \vec{r}'|\right)d\eta' = -e^{-ikx\cos\varphi_0}\cos(ky\sin\varphi_0). \qquad (2.26b)$$

We again use relations (2.22) and (2.23) and the fact that the functions ce_n and se_n are orthogonal on the interval $[0, \pi]$ and obtain

$$I^e(\eta) = \frac{-i}{f\sqrt{1 - \cos^2\eta}}\sum_{n=0}^\infty c_n\frac{Ce_n(\xi_1, q)}{Me_n^{(2)}(\xi_1, q)}Me_n^{(2)'}(0, q)ce_n(\varphi_0, q)ce_n(\eta, q),$$

$$\qquad (2.27)$$

$$I^m(\eta) = i\sum_{n=0}^\infty d_n\frac{Se_n(\xi_1, q)}{Ne_n^{(2)}(\xi_1, q)}Ne_n^{(2)}(0, q)se_n(\varphi_0, q)se_n(\eta, q). \qquad (2.28)$$

The same result can be obtained using representation (1.2a), where an ellipse confocal with S but with lesser axes is first taken as Σ and then contracted to an interval $|x| \leq f$ [9, Chapter 2]. Substituting (2.27) and (2.28) into (2.25), we again obtain the series (2.24) for the diffraction field.

Solving the same problem with the Neumann boundary condition $\frac{\partial}{\partial n}(U^0 + U^1)|_S = 0$, we can similarly obtain [9, Chapter 2]

$$I_N^e(\eta) = \frac{-i}{f\sqrt{1 - \cos^2\eta}}\sum_{n=0}^\infty c_n\frac{Ce_n'(\xi_1, q)}{Me_n^{(2)'}(\xi_1, q)}Me_n^{(2)'}(0, q)ce_n(\varphi_0, q)ce_n(\eta, q),$$

$$\qquad (2.29)$$

$$I^m(\eta) = i\sum_{n=0}^\infty d_n\frac{Se_n'(\xi_1, q)}{Ne_n^{(2)'}(\xi_1, q)}Ne_n^{(2)}(0, q)se_n(\varphi_0, q)se_n(\eta, q). \qquad (2.30)$$

It follows from the obtained relations that the equivalent electric current distributed over the interval $-f \leq x \leq f$, $y = 0$, has singularities near the ends of this interval (i.e., for $\eta = 0$, $\eta = \pi$) of the form $[(f - x)(f + x)]^{-1/2}$ for both of the boundary conditions, which agrees well with the result obtained in Section 1.2.5. We also note that obtained rigorous relations allow us clearly to verify that the integral equations of MAC have solutions for a correct (according to the existence theorem) choice of the auxiliary contour.

When the MAC integral equations are solved numerically, the support of auxiliary currents must also be chosen with the requirements of the existence theorem taken into account. As previously noted, one of the most popular MAC versions is the MDS. The main concept of this method was discussed above (see Section 2.2). The next section deals with a modification of both methods (MAC and MDS), where the support of auxiliary currents is constructed by a uniform and efficient procedure.

2.4 MODIFIED MDS

The size N of the algebraic system in MDS is determined, in particular, by the required accuracy of the solution of the initial boundary-value problem. As N increases, the sources on Σ move closer together, and therefore, it is also required in MDS that the support Σ surrounds the singularities of the field $U^1(\vec{r})$. However, in other respects, the choice of the contour Σ is sufficiently ambiguous. At the same time, the rate of the calculation algorithm convergence significantly depends on the configuration of Σ, but there is not a unique rule for choosing this configuration. The support Σ is often chosen as a contour similar to S [25, Chapter 8], but such a choice is not optimal in the majority of cases.

Consider one example of solving the two-dimensional problem of wave scattering at a smooth body by the MDS is used to show that the contour Σ must be constructed by an analytic deformation of the contour S [26]. In this case, the least discrepancy $\min \Delta$ of the boundary condition satisfaction at the points of the contour S between the collocation points can be attained if the contour Σ is "densely spanned" by the singularities of the wave field $U^1(\vec{r})$ [26]. Following this [26], we rewrite (2.12) as

$$|\vec{r}_S - \vec{r}_\Sigma| = \left[(z - \zeta)(\bar{z} - \bar{\zeta}) \right]^{1/2},$$

where $z = re^{i\alpha}$, $\alpha \in [0, 2\pi]$,

$$\zeta = \rho e^{i\theta}, \ \theta = \theta' + i\theta'', \ \theta' \in [0, 2\pi]. \tag{2.31}$$

If $\theta'' = 0$ in (2.31), then the variable $\zeta \in C$, where C is the contour on the complex plane z corresponding to S under the mapping (2.31) and congruent to it. However, if the parameter θ is assumed to be complex, i.e., if we assume that $\theta'' \neq 0$, then the contour C experiences a deformation, in particular, it contracts for positive θ''. By setting

$$r_\Sigma = |\zeta|, \quad \phi_\Sigma = \arg\zeta, \tag{2.32}$$

we then determine the required geometry of the contour Σ and take this contour as the support of the auxiliary current density. Such a deformation is possible until the mapping (2.31) ceases to be one-to-one. The points at which this mutual uniqueness is violated satisfy the relation

$$\zeta' \equiv [\rho'(\theta) + i\rho(\theta)]e^{i\theta} = 0, \tag{2.33}$$

and we consider only the roots that correspond to points on the complex plane z inside the contour C: $\zeta(\theta)$, $\theta \in [0, 2\pi]$. Precisely these points, as well as the singular points of the function $U^0(\vec{r})|_S$ under its analytic continuation into a domain inside S form the set of singularities of the diffraction field $U^1(\vec{r})$ (see Section 1.2.2). For Equation (2.2) to be solvable, it is necessary and sufficient to surround these singularities by the contour Σ.

In the case of figures such as ellipse, Cassinian oval, and multifoil, the roots of Equations (2.33) have simple analytic expressions as shown above (see Section 1.2.3.1), and therefore, an appropriate θ'' can easily be calculated. In more difficult situations, θ'' can be obtained numerically [59, 60]. As in Kyurkchan et al. (2001) [26], we characterize the position of the contour Σ by a parameter $\delta \in]0, 1[$, where 0 corresponds to the case, where Σ passes through singular points, and 1 corresponds to the case, where Σ coincides with S.

The previous study showed that the construction of an auxiliary contour Σ by the above-described analytic deformation of the contour S ensures stability and a high accuracy of the numerical algorithm. At the same time, when choosing a contour Σ similar or equidistant to S, one meets the problem of stability and accuracy of the computational algorithm, because the matrix of the corresponding algebraic system turns out to be almost degenerate when one attempts to obtain an appropriate accuracy.

Figure 2.2 illustrates the trajectories of motion of points from the four-leaf initial contour S to the auxiliary contours such as Σ_1, similar to the initial contour S, and Σ_2 obtained by its analytic deformation [57, 58]. One can see that the points move in the shortest way in the last case. As a result, the matrices of the algebraic systems corresponding to the contour analytic deformation have a pronounced principal diagonal and, as a consequence, are well conditioned.

In the case of the Neumann boundary condition, the support Σ is chosen according to the technique described above. From the technical standpoint,

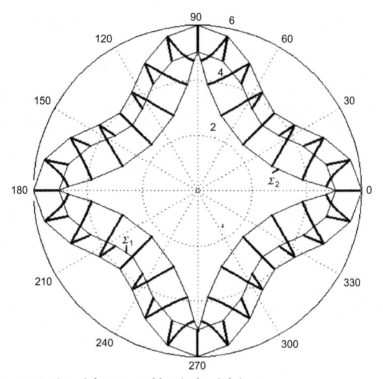

Figure 2.2 Analytic deformation of four-leaf multifoil.

it is expedient to consider a fictitious "electric current" whose integral equation can be written as [26]

$$\frac{\partial}{\partial n_S} U^0\left(\vec{r}_S\right) = \int_\Sigma I^E\left(\vec{r}_\Sigma\right) \frac{\partial}{\partial n_S} H_0^{(2)}\left(k|\vec{r}_S - \vec{r}_\Sigma|\right) d\sigma, \quad \vec{r}_S \in S. \qquad (2.34)$$

The above existence theorem also holds in this case.

Let us illustrate the above technique by an example of solving the diffraction problem for the plane wave $U^0\left(\vec{r}\right) = \exp\{-ikr\cos\left(\varphi - \varphi_0\right)\}$, where φ_0 is the direction of the wave propagation with respect to the axis Ox, scattered at a cylinder with the multifoil-shaped directrix $r(\alpha) = a(1 + \tau\cos q\alpha)$, $0 \leq \tau < 1, q = 2, 3, \ldots$, and with the boundary condition (2.1) satisfied on its surface. The singularities of the diffraction field in this problem are listed above (see (1.71) and (1.72)).

We let $\Delta_{m+1/2}$ denote the sum of the left- and right-hand sides of the algebraic system (2.17), where m is replaced by $m + 1/2$. Then

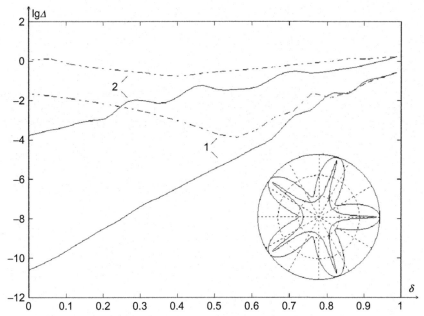

Figure 2.3 Boundary condition discrepancy for five-leaf multifoil.

$\Delta = \max_m |\Delta_{m+1/2}|$ is the measure of the error of the obtained solution on the boundary.

Figure 2.3 illustrates how the discrepancy $\lg \Delta$ depends on δ for different deformation parameters τ. Here δ determines the position of the auxiliary contour between the lobe vertex and the corresponding singular point. The auxiliary contour shown in the insert in Fig. 2.3 was obtained by an analytic deformation of the initial multifoil contour and corresponds to $\delta = 0,01$. The results of calculations are depicted by solid lines. The dash-dotted lines illustrate the results obtained in the case, where Σ is chosen as a contour similar to S. Here the parameter δ varies from 0.99 to 0.01. The calculations were performed for the pentafoil with $ka = 10$. The number of equations in system (2.17) was equal to 128. Digit 1 denotes the curves corresponding to $\tau = 0.1$, digit 2 corresponds to $\tau = 0.5$. One can see that in the case, where Σ is constructed by an analytic deformation of S, the discrepancy monotonically decreases with decreasing δ. To obtain the desired accuracy of calculations in more complicated situations, it is necessary to "surround" the set of singularities by the contour Σ more closely, i.e., to take even lesser values of the parameter δ [59, 60].

Table 2.1 Maximum discrepancy values depending on the distance from the auxiliary contour to singular points.

δ	10^{-3}	10^{-4}	10^{-5}	10^{-6}	10^{-7}
$N=128$ max(Δ)	0.01256	0.0005825	0.0005807	0.00058057	0.00058055
$N=256$ max(Δ)	6.9×10^{-8}	5.03×10^{-9}	3.54×10^{-10}	8.71×10^{-11}	6.44×10^{-11}

In Table 2.1, we illustrate the last assertion by presenting the results of calculations [58, 59] of max(Δ) in the problem of E-polarized plane wave scattering at an dielectric cylinder with a pentafoil cross-section for the following values of the problem parameters: $ka = 10$; $\tau = 0,2$; $\varepsilon_i = 7$.

In the case of wave scattering at a dielectric cylinder with relative dielectric permittivity ε_i, the initial scattering problem can be reduced to solving the system of integral equations

$$U^0\left(\vec{r}_S\right) + \int_{\Sigma} I\left(\vec{r}_\Sigma\right) H_0^{(2)}\left(k|\vec{r}_S - \vec{r}_\Sigma|\right) d\sigma = \int_{\Sigma_1} I_1\left(\vec{r}_{\Sigma_1}\right) H_0^{(2)}\left(k_i|\vec{r}_S - \vec{r}_{\Sigma_1}|\right) d\sigma_1,$$

$$\frac{\partial}{\partial n_S}\left[U^0\left(\vec{r}_S\right) + \int_{\Sigma} I\left(\vec{r}_\Sigma\right) H_0^{(2)}\left(k|\vec{r}_S - \vec{r}_\Sigma|\right) d\sigma\right] =$$

$$\chi \frac{\partial}{\partial n_S} \int_{\Sigma_1} I_1\left(\vec{r}_{\Sigma_1}\right) H_0^{(2)}\left(k_i|\vec{r}_S - \vec{r}_{\Sigma_1}|\right) d\sigma_1,$$

$$(2.35)$$

where $\chi = 1$ for the E-polarized incident wave and $\chi = 1/\varepsilon_i$ for the H-polarized incident wave; Σ, Σ_1 are the contours on which the equivalent currents $I\left(\vec{r}_\Sigma\right)$, $I_1\left(\vec{r}_{\Sigma_1}\right)$ are located. System (2.35) is solvable if, and only if, the contour Σ surrounds the singularities of the continuation of the solution U^1 of the external boundary-value problem into a domain inside S and the contour Σ_1 surrounds the singularities of the continuation of the solution U^i of the internal boundary-value problem outside S. In this example, the contour Σ_1 is also constructed by an analytic deformation of the boundary S.

Now let us consider other examples of application of the approach discussed above. We consider the problem of diffraction of a plane monochromatic E-polarized wave

$$U^0 = \exp\{-ikr \cos\left(\varphi - \varphi_0\right)\}, \qquad (2.36)$$

at an infinite elliptic cylinder with the directrix $\rho(\varphi) = b/\sqrt{1 - \varepsilon^2 \cos^2\varphi}$ and semiaxes a and b, $a > b$, and with the homogeneous Dirichlet condition for the total field satisfied on its surface. It was previously shown that the singularities of the field scattered at an ellipse when the plane wave is incident on it are located at its foci. The calculations were performed for $kb = 0.1$, $ka = \sqrt{28}$, and $\varphi_0 = \pi/2$. For the number of basic functions $M = 256$ and for $\delta = 10^{-6}$, the calculated scattering pattern $g(\alpha)$ coincides with the graphical accuracy with the scattering pattern of a band of half-width $ka = \sqrt{28}$ [132]. In this case, the maximal discrepancy of the boundary condition (max Δ) was equal to 0.4, and it was observed in a small neighborhood of the angles $\varphi = 0, \varphi = \pi$, i.e., near the points of the contour S that are the closest to the field singularities. At the other points of the contour, Δ was of the order of 0.01. In the case of H-polarization (the Neumann boundary condition), the results of calculations coincide with the results obtained in Ref. [132]. It should be noted that we do not know the works, where the methods of auxiliary currents or discrete sources were used to solve the problem of diffraction at such an oblong ellipse ($a/b = 53$). Our attempts to obtain the solution by choosing Σ as a contour similar to S remained unsuccessful.

Figure 2.4 shows the results of calculations of the discrepancy Δ for an elliptic cylinder with semiaxes $kb = 1$, $ka = 5$ ($M = 128$, $\delta = 10^{-6}$) in the case where the field of the incident wave $U^0(\vec{r})$ is the field of an electric current filament (cylindrical wave) [59, 60]

$$U^0(\vec{r}) = H_0^{(2)}(k|\vec{r} - \vec{r}_0|), \tag{2.37}$$

where $\vec{r}_0 = \{r_0, \varphi_0\}$ are the source coordinates (we set $\varphi_0 = \pi/2$ in the calculations whose results are given below) for different distances r_0 from the

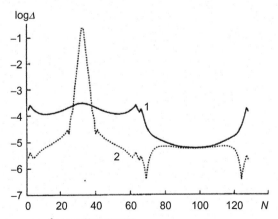

Figure 2.4 Discrepancy for a point source.

source to the origin. We note that the situation in this case is more compli-cated than in the case of scattering of the plane wave (2.36). The point is that as the source approaches the body surface, i.e., as the condition $r_0 < 2a/\varepsilon$ begins to be satisfied, the source image (2.37) ("imaginary" source) moves from the nonphysical sheet of the Riemann surface to the principal (physical) sheet (see Section 1.2.3.2). Therefore, according to the above-cited exis-tence theorem, the contour Σ must also surround the source image. In Fig. 2.4, curve 1 corresponds to the case $kr_0 = 1.5$, and curve 2 corresponds to the case $kr_0 = 1.01$. One can see that for $kr_0 = 1.5$, the discrepancy is approximately equal to 10^{-4} on the average. For $kr_0 = 1.01$, the source image is practically at the "mirror point" (see Section 1.2.3.2) inside S under the source. The contour Σ does not surround the image.

As a result, the discrepancy value is sufficiently large $(\max(\Delta) \approx 0.5)$ in a small neighborhood of the contour S point directly under the source. How-ever, on the rest of S, the value of Δ does not exceed 10^{-5}, and therefore, such a local overshoot of the boundary condition discrepancy does not prac-tically affect the results when the integral characteristics (scattering patterns, effective scattering diameter, etc.) are calculated.

Figure 2.5 illustrates the results of calculations of Δ for $kr_0 = 1.5$ by the standard method of discrete sources (SMDS) (where the contour Σ is similar to S). The value of δ is then taken to be equal to 0.25, which corresponds to the optimal position of Σ in SMDS [26]). One can see that Δ is here of the order of unity. For $kr_0 = 1.01$, the SMDS algorithm is generally destroyed. Thus, when using the SMDS technique, it is necessary to distinguish the contributions to the diffraction field made by the image and by the "cut" connecting the image with the contour Σ. However, as previously shown, this is unnecessary when

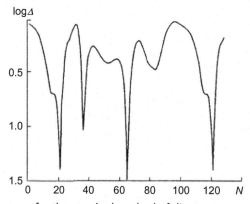

Figure 2.5 Discrepancy for the standard method of discrete sources.

the MDS modification discussed here is used. We believe that this fact is rather important, because it allows one to use a unified technique independent of the form of the incident field and the boundary conditions.

The above-discussed MMDS scheme was successfully used to solve rather many problems. For example, a further development of MMDS has been proposed [52], which allows one to solve the diffraction problems in the situations where, in addition to the singularities of the mapping (2.31), it is also necessary to consider other singular points. Such singularities can arise not only in the case of the scatterer radiation by the field of a point source, but also in the case of diffraction at a group of closely located bodies, for example, in the situation similar to that shown in Fig. 1.5. In such a situation, inside the scatterer, there can also be singularities "induced" by the field of the singular points lying inside another scatterer located at a short distance from the first one.

Now we consider the problem of normal incidence of a plane monochromatic wave on an infinite surface with unbounded cross-section of sinusoidal shape $y = a\cos px$. Figure 2.6 illustrates the results of computations for the following parameters of the problem: $b/\lambda = 2.5$, curves 1 correspond to $ka = 2.5$, $N = 16$, and curves 2 correspond to $ka = 25$, $N = 128$. One can

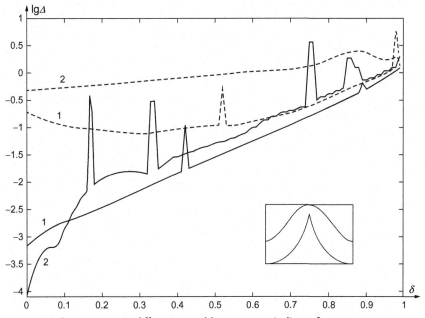

Figure 2.6 Discrepancy in diffraction problem on a periodic surface.

see that the best results are attained in MMDS when the auxiliary contour is closely pressed to the point y_0^1 determined by (1.74), which corresponds to $\delta \approx 0$.

There are examples of MMDS generalizations to three-dimensional and vector problems [71, 74, 76, 81, 82].

To conclude this section, we note that if it is required to solve the problem of diffraction at a body whose shape strongly differs from a circle or a sphere (e.g., strongly oblate or strongly oblong body), then using MMDS (or MMAC), it is expedient to pass from the spherical (or cylindrical) coordinates to the coordinates more adequate to the geometry of the body in question. The corresponding approach is discussed in detail in Ref. [133].

CHAPTER 3

Null Field and *T*-Matrix Methods

The mathematical models of wave scattering and diffraction phenomena are based on a wide range of numerical methods. Most of the rigorous methods, i.e., the methods that principally permit calculating in the framework of a chosen model with a prescribed accuracy, employ some analytic representations of the wave field. Further, the use of the boundary condition allows one to reduce the problem to either an algebraic system for the expansion coefficients or an integral or integro-differential equation for the distribution density of some sources. It was shown in the preceding chapter that such methods lead to correct computational algorithms only if information about the analytic properties of the solution is used.

The null field method (NFM) [5, 8, 19] and one of its modifications, the *T*-matrix method, are most widely used to solve the wave diffraction problems [20, 28, 29]. However, no information about the singularities of the analytic continuation of the diffraction field was used either in Waterman's pioneer paper [19] or in the subsequent works (e.g., see [20, 28, 29] and the references therein). As shown in Section 2.1 (also see [9, 10]), the condition that the support of auxiliary currents surrounds the set of singularities of the wave field analytic continuation is necessary and sufficient for the integral equation of the method of auxiliary currents (MAC) to be solved. Below we show that this condition is also of principal importance in NFM.

3.1 NFM FOR SCALAR DIFFRACTION PROBLEMS

3.1.1 Statement of the Problem and Derivation of the NFM Integral Equation

First, we consider the scalar problem of wave diffraction at a compact scatterer occupying a spatial domain D with the boundary S. In the domain $D_e = \mathbb{R}^3 \setminus \bar{D} (D_e = \mathbb{R}^2 \setminus \bar{D})$, it is required to find the solution $U^1(\vec{r})$ of the Helmholtz equation

$$\Delta U^1(\vec{r}) + k^2 U^1(\vec{r}) = 0 \qquad (3.1)$$

Mathematical Modeling in Diffraction Theory
http://dx.doi.org/10.1016/B978-0-12-803728-7.00003-8

(where $k = \omega\sqrt{\varepsilon\mu}$ is the wave number, $k \in C$, $k \neq 0$), which satisfies some boundary conditions on the boundary S of the domain D; for definiteness, this can be, for example, the boundary condition of the third type

$$\left[U(\vec{r}) - \frac{Z}{ik\zeta} \frac{\partial U(\vec{r})}{\partial n} \right]\Bigg|_S = 0, \tag{3.2}$$

where Z is the impedance of the surface S, ζ is the wave resistance of the free space, and $\partial/\partial n$ is the differentiation in the direction of the outer normal on S; the solution must also satisfy the Sommerfeld radiation condition at infinity

$$\lim_{r \to \infty} r^{(n-1)/2}\left(\frac{\partial U^1}{\partial r} + ikU^1 \right) = 0. \tag{3.3}$$

In Section 1.1.1, it was shown (see (1.9)) that

$$\int_S \left\{ U(\vec{r}') \frac{\partial G_0(\vec{r}, \vec{r}')}{\partial n'} - \frac{\partial U(\vec{r}')}{\partial n'} G_0(\vec{r}, \vec{r}') \right\} ds' + U^0(\vec{r}) = 0, \quad M(\vec{r}) \in D. \tag{3.4}$$

Formula (3.4) (and (1.6)) is called the null field condition. This condition allows one to obtain a Fredholm integral equation of the first type with a smooth kernel, and precisely this equation underlies the null field and T-matrix methods. We note that the word combination "null field condition" does not completely explain the meaning of formulas (3.4) and (1.6), because the integral on the left-hand sides of these relations is not the wave field inside S. However, some representations of the boundary values of the quantities $U|_S$ and $\partial U/\partial n|_S$ must be used in any attempt to solve these equations, and already in this case, for example, (3.4) means that the sum of the diffraction and incident fields is zero (see below). The null field condition (3.4) is satisfied everywhere inside the scatterer, but as we see below, for the correct application of this condition when solving the boundary-value problems, it is necessary to have the complete information about the singularities of the wave field analytic continuation into the interior of the scatterer.

We consider the case where the surface S is analytic, i.e., where it can be prescribed by an equation analytically depending on the variables.

To obtain the null field integral equation, we assume that condition (3.4) is satisfied on a closed surface Σ inside S and take the boundary condition (3.2) into account. As a result, we obtain the Fredholm integral equation of the first type with a smooth kernel

$$U^0\left(\vec{r}\right) + \int_S \frac{\partial U\left(\vec{r}'\right)}{\partial n'}\left\{\frac{Z}{ik\zeta}\frac{\partial G_0\left(\vec{r},\vec{r}'\right)}{\partial n'} - G_0\left(\vec{r},\vec{r}'\right)\right\}ds' = 0, \quad M\left(\vec{r}\right) \in \Sigma.$$

(3.5)

If the internal field U^i is nonzero, i.e., if the conjugation boundary conditions are satisfied,

$$\begin{cases} U\left(\vec{r}\right)\big|_S = U^i\left(\vec{r}\right)\big|_S, \\ \dfrac{1}{\rho}\dfrac{\partial U\left(\vec{r}\right)}{\partial n}\bigg|_S = \dfrac{1}{\rho_i}\dfrac{\partial U^i\left(\vec{r}\right)}{\partial n}\bigg|_S, \end{cases}$$

(3.6)

where ρ, ρ_i are parameters respectively characterizing the optical density outside and inside S, then to obtain the integral equations, it is also necessary to use the null field condition for U^i (the upper row in (1.8)). Let us find the null field conditions for U and U^i on the surfaces Σ_1 and Σ_2, respectively, which are chosen so that the surface Σ_1 is inside S and the surface Σ_2 is outside S. With the boundary conditions (3.6) taken into account, we obtain the system of two Fredholm integral equations of the first type

$$\begin{cases} \displaystyle\iint_S \left\{U\left(\vec{r}'\right)\frac{\partial G_0\left(\vec{r},\vec{r}'\right)}{\partial n'} - \frac{\partial U\left(\vec{r}'\right)}{\partial n'}G_0\left(\vec{r},\vec{r}'\right)\right\}ds' + U^0\left(\vec{r}\right) = 0, \quad M\left(\vec{r}\right)\in\Sigma_1, \\ \displaystyle\iint_S \left\{U\left(\vec{r}'\right)\frac{\partial G_i\left(\vec{r},\vec{r}'\right)}{\partial n'} - \frac{\rho_i}{\rho}\frac{\partial U\left(\vec{r}'\right)}{\partial n'}G_i\left(\vec{r},\vec{r}'\right)\right\}ds' = 0, \quad M\left(\vec{r}\right) \in \Sigma_2. \end{cases}$$

(3.7)

Equation (3.4) has a solution corresponding to the above-posed boundary-value problem if, and only if, the surface Σ surrounds the set $\bar{A}$ of singularities of the analytic continuation of the field scattered at a body bounded by the surface S. Indeed, to solve equations (3.4), it is necessary to use some analytic representations of the scattered field $U^1\left(\vec{r}\right)$. But all such representations are meaningful only outside the set $\bar{A}$. Moreover, we previously showed [22, 23, 78–80] that the most stable and fast-operating algorithms are implemented when the auxiliary surface is constructed by an analytic deformation of the scatterer boundary to the wave field singularities, just as in the modified MAC. Obviously, this reasoning also holds for system (3.7), which has a solution corresponding to the above-posed boundary-value problem if and only if the surface Σ_1 surrounds the set of singularities of the analytic continuation of the field U^1 into the interior of S and the surface Σ_2 surrounds the set of singularities of the analytic continuation of the field U^i into the environment.

By analogy with the terminology used in Chapter 2, the NFM, where the null field condition is satisfied on a surface obtained by an analytic deformation of the scatterer boundary, will be called the modified null field method (MNFM).

3.1.2 Numerical Solution of the NFM Integral Equation

As in Chapter 2, we solve integral Equation (3.5) by using the well-known Krylov–Bogolyubov technique [11, 43], which states that, to approximate the unknown function in the integrand, it is necessary to use a basis of characteristics subintervals obtained by dividing the interval of integration (a piecewise constant basis) and to equate the left- and right-hand sides of the integral equations at the collocation points. The corresponding algorithm is considered in Chapter 2 for two-dimensional problems, but in this chapter, we mostly consider three-dimensional problems in the case where S is the surface of a body of revolution, and hence we here present the required computational algorithm.

1. We pass to spherical coordinates (r, θ, φ), where $r' = \rho(\theta')$ is the equation of the surface S, $r = r(\theta)$ is the equation of the surface Σ,

$$ds' = \kappa(\theta')\rho(\theta')d\theta'd\varphi', \kappa(\theta') = \sqrt{\rho^2(\theta') + \rho'^2(\theta')}, G_0 = \frac{1}{4\pi} \cdot \frac{e^{-ikR(\theta',\theta,\varphi-\varphi')}}{R(\theta',\theta,\varphi-\varphi')},$$

$$\frac{\partial G_0(\vec{r},\vec{r'})}{\partial n'} = \frac{1}{\kappa(\theta')}\left(r(\theta')\frac{\partial G_0(\vec{r},\vec{r'})}{\partial r'} - \frac{r'(\theta')}{r(\theta')}\frac{\partial G_0}{\partial \theta'}\right),$$

$$\frac{\partial G_0}{\partial r'} = \frac{\partial G_0}{\partial R}\frac{r - \rho\cos\gamma}{R}, \frac{\partial G_0}{\partial \theta'} = \frac{\partial G_0}{\partial R}\frac{-r\rho\frac{\partial\cos\gamma}{\partial\theta'}}{R}, \frac{\partial G_0}{\partial R} = -G_0\left(ik + \frac{1}{R}\right),$$

$$R(\theta',\theta,\varphi-\varphi') = \sqrt{\rho^2(\theta') + r^2(\theta) - 2\rho(\theta')r(\theta)\cos\gamma},$$

$$\cos\gamma = \sin\theta\sin\theta'\cos(\varphi-\varphi') + \cos\theta\cos\theta'.$$

As a result, Equation (3.5) becomes

$$\int_0^{2\pi}\int_0^{\pi} I(\theta',\varphi') \cdot K(\theta,\theta',t)d\theta'd\varphi' = -U^0(\theta,\varphi), \quad M(\theta',\varphi') \in S, M(\theta,\varphi) \in \Sigma,$$

$$(3.8)$$

where

$$\frac{\partial U(\vec{r}'')}{\partial n'} = \frac{1}{\kappa(\theta')r(\theta')\sin(\theta')} I(\theta',\varphi'),$$

$$K(\theta,\theta',t) = -G_0(\theta,\theta',t)$$

$$\times \left\{ 1 + \frac{Z}{ik\zeta} \cdot \left(\frac{i}{R} + \frac{1}{kR^2} \right) \frac{1}{\kappa(\theta')} \left(r^2(\theta') - r(\theta')\rho(\theta)\cos\gamma + r'(\theta')\rho(\theta)\frac{\partial\cos\gamma}{\partial\theta'} \right) \right\}.$$

2. We use the rotational symmetry to expand the functions $U^0(\theta,\varphi)$, $I(\theta',\varphi')$, and $K(\theta,\theta',t)$ in the Fourier series in φ, φ', and $t = \varphi - \varphi'$, respectively:

$$U^0(\theta,\varphi) = \sum_{s=-\infty}^{\infty} U_s^0(\theta) \cdot e^{i \cdot s \cdot \varphi},$$

$$I(\theta',\varphi') = \sum_{n=-\infty}^{\infty} I_n(\theta') \cdot e^{i \cdot n \cdot \varphi'}, \tag{3.9}$$

$$K(\theta,\theta',t) = \sum_{s=-\infty}^{\infty} K_s(\theta,\theta') \cdot e^{i \cdot s \cdot t}. \tag{3.10}$$

The coefficients of the Fourier expansion of the kernel of Equation (3.8) are

$$K_s(\theta,\theta') = \frac{1}{8\pi^2} \int_0^{2\pi} K(\theta,\theta',t) \cdot e^{-i \cdot s \cdot t} dt. \tag{3.11}$$

The direct calculation of the Fourier coefficients of Green's function and its derivatives (3.11), which are known in the literature as Vasiliev *S*-functions [12], is a rather difficult problem, but in practice, it is efficient in the case where it suffices to take a number of Fourier harmonics that is not too large and simply to replace the integrals (3.11) by sums according to the rectangle rule

$$K_s(\theta,\theta') \cong \frac{1}{4\pi N_1} \sum_{\nu=1}^{N_1} K(\theta,\theta',t_\nu) e^{i \cdot s \cdot t_\nu}, \tag{3.12}$$

where

$$t_\nu = \frac{2\pi(\nu - 0.5)}{N_1}.$$

3. Further, after simple transformations, we truncate these series by taking finite sums and obtain $2Q + 1$ integral equations:

$$2\pi \int_0^\pi I_s(\theta') \cdot K_s(\theta, \theta') d\theta' = -U_s^0(\theta), \quad M(\theta') \in S, \ M(\theta) \in \Sigma, \quad (3.13)$$

where $K_s(\theta, \theta')$ is calculated by Equation (3.12),

$$U_s^0(\theta) = \frac{1}{2\pi} \int_0^{2\pi} U^0(\theta, \varphi) \cdot e^{-i \cdot s \cdot \varphi} d\varphi, \quad s = \overline{-Q, Q}.$$

4. Now we perform a piecewise constant approximation of the Fourier harmonics of the unknown function $I_s(\theta')$:

$$I_s(\theta') = \sum_{n=1}^N I_{n,s} \cdot \phi_n(\theta'), \quad (3.14)$$

where $\phi_n(\theta') = \begin{cases} 1, \theta' \in [(n-1) \cdot \pi/N, n \cdot \pi/N], \\ 0, \theta' \notin [(n-1) \cdot \pi/N, n \cdot \pi/N]. \end{cases}$

5. Then we apply the collocation method. We divide the interval $[0, \pi]$ of θ variation into $N \geq N_1$ equal parts, write Equations (3.13) at the collocation points $\theta = \theta_m = \frac{\pi}{N} \cdot (m - 0.5)$, $m = \overline{1, N}$, coinciding with the middle points of the subintervals, and obtain $2Q + 1$ systems of N linear algebraic equations (SLAE)

$$2\pi \sum_{n=1}^N I_{n,s} \int_{(n-1)\pi/N}^{n\pi/N} K_s(\theta_m, \theta') d\theta' = -U_s^0(\theta_m), \quad M(\theta') \in S, \ M(\theta) \in \Sigma.$$

$$(3.15)$$

6. Next, we can solve SLAE (3.15) directly by using some known quadrature formula to calculate the integrals contained in it, or we can pass to discrete sources, which is equivalent to replacing the integrals in (3.15) by sums of the integrand values at the middle points of the integration intervals multiplied by the length of the interval:

$$\frac{2\pi^2}{N} \sum_{n=1}^N I_{n,s} K_s(\theta_m, \theta_n) = -U_s^0(\theta_m), \quad M(\theta') \in S, \ M(\theta) \in \Sigma. \quad (3.16)$$

3.2 NFM FOR VECTOR DIFFRACTION PROBLEMS

3.2.1 Statement of the Problem and Derivation of the NFM Integral Equation

Now we consider the vector case. It is required to find the vectors $\vec{E}^1, \vec{H}^1$ which, in $\mathbb{R}^3 \backslash \bar{D}$, satisfy the homogeneous system of Maxwell equations

$$\nabla \times \vec{E}^1 = -ik\zeta \vec{H}^1, \quad \nabla \times \vec{H}^1 = \frac{ik}{\zeta}\vec{E}^1, \tag{3.17}$$

the Sommerfeld condition at infinity, and the boundary condition

$$\left(\vec{n} \times \vec{E}\right)\big|_S = Z\left(\vec{n} \times \left(\vec{n} \times \vec{H}\right)\right)\big|_S, \tag{3.18}$$

where $\vec{E} = \vec{E}^0 + \vec{E}^1$, $\vec{H} = \vec{H}^0 + \vec{H}^1$, and $\vec{E}^0, \vec{H}^0$ is the initial (incident) field, and Z is the surface impedance.

For the problem in question, the null field condition has the form (see (1.24))

$$\int_S \left\{ \frac{\zeta}{ik}\left(k^2 \vec{J} G_0(\vec{r}, \vec{r}') - \left(\vec{J} \cdot \nabla'\right)\nabla G_0\right) + \vec{J}^m \times \nabla G_0 \right\} ds' = -\vec{E}^0(\vec{r}),$$

$$M(\vec{r}) \in D, \tag{3.19}$$

where $\vec{J} = \left(\vec{n} \times \vec{H}\right)\big|_S$, $\vec{J}^m = \left(\vec{E} \times \vec{n}\right)\big|_S$.

Satisfying condition (3.19) on a surface Σ inside S and taking the boundary conditions into account, we obtain the integral equation

$$\left\{ \vec{n} \times \frac{\zeta}{4\pi ik}\int_S \left\{ \left[k^2 \vec{J}(\vec{r}')G_0(\vec{r};\vec{r}') - \left(\vec{J} \cdot \nabla'\right)\nabla G_0(\vec{r};\vec{r}')\right] \right.\right.$$

$$\left.\left. - Z\left(\vec{n}' \times \vec{J}\right)\nabla G_0 \right\} ds' \right\}\bigg|_\Sigma = -\left(\vec{n} \times \vec{E}^0(\vec{r})\right)\bigg|_\Sigma, \tag{3.20}$$

As in the scalar case, Equation (3.20) has a solution corresponding to the above-posed boundary-value problem (3.17) and (3.18) only if the surface Σ surrounds the set $\bar{A}$ of singularities of the wave field continuation.

We use the algorithm considered in Section 3.1.2 to solve integral Equation (3.20) numerically.

3.3 RESULTS OF NUMERICAL STUDIES

To estimate the reliability of the numerical solution of integral equations, we calculated the scattering pattern (see Sections 1.1.3 and 1.2.1), some of whose properties are already known from the statement of the problem. Moreover, for some model problems, the pattern can be compared with the patterns obtained by other numerical methods (and even by analytical methods in some cases).

For scalar diffraction problems in the case of the boundary condition of the third type, the scattering pattern has the form

$$f(\theta, \varphi) = \int_0^{2\pi} \int_0^{\pi} I(\theta', \varphi') e^{ikr(\theta')\cos\gamma} \left[1 + \frac{Z}{k\zeta} \cdot \frac{1}{\kappa(\theta')} \left(r(\theta')\cos\gamma - r'(\theta') \frac{\partial\cos\gamma}{\partial\theta} \right) \right] d\theta' \, d\varphi',$$

(3.21)

and substituting the expansions of the function $I(\theta', \varphi')$ and replacing the integrals in (3.21) by the Riemann sums, we obtain

$$f(\theta, \varphi) = -\frac{2\pi^2}{N} \sum_{s=-Q}^{Q} i^s e^{is\varphi} \sum_{n=1}^{N} I_{n,s} e^{ikr(\theta_n)\cos\theta_n\cos\theta}$$

$$\times \left[J_s(kr(\theta_n)\sin\theta\sin\theta_n) \left(1 + \frac{Z}{k\zeta}\frac{1}{\kappa(\theta_n)}(-\cos\theta(r(\theta_n)\cos\theta_n + r'(\theta_n)\sin\theta_n)) \right) \right.$$

$$+ \frac{Z}{ik\zeta}\frac{1}{2\kappa(\theta_n)}(-J_{s+1}(kr(\theta_n)\sin\theta\sin\theta_n) + J_{s-1}(kr(\theta_n)\sin\theta\sin\theta_n))$$

$$\left. \times (\sin\theta(-r(\theta_n)\sin\theta_n + r'(\theta_n)\cos\theta_n)) \right].$$

For the vector problem in the case of bodies of revolution, the components of the scattering pattern have the form

$$F_\theta^E(\theta, \varphi) = \frac{k^2\zeta}{4\pi i} \int_0^{2\pi} \int_0^{\pi} \left\{ J_1(\alpha, \beta) \left[\frac{\rho'(\alpha)}{\rho(\alpha)} \frac{\partial\cos\gamma}{\partial\theta} + \frac{\partial^2\cos\gamma}{\partial\theta\partial\alpha} \right] \right.$$

$$\left. + \frac{J_2(\alpha, \beta)}{\sin\alpha} \frac{\partial^2\cos\gamma}{\partial\theta\partial\beta} \right\} e^{ik\rho\cos\gamma} \rho(\alpha) d\alpha d\beta$$

$$F_\varphi^E(\theta, \varphi) = \frac{k^2\zeta}{4\pi i} \int_0^{2\pi} \int_0^{\pi} \left\{ J_1(\alpha, \beta) \left[\frac{\rho'(\alpha)}{\rho(\alpha)} \frac{\partial\cos\gamma}{\partial\varphi} + \frac{\partial^2\cos\gamma}{\partial\varphi\partial\alpha} \right] \right.$$

$$\left. + \frac{J_2(\alpha, \beta)}{\sin\alpha} \frac{\partial^2\cos\gamma}{\partial\varphi\partial\beta} \right\} e^{ik\rho\cos\gamma} \rho(\alpha) d\alpha d\beta,$$

where $\vec{J} = \vec{i}_\rho \frac{\rho'(\alpha)}{\rho(\alpha)} J_1 + \vec{i}_\alpha J_1 + \vec{i}_\beta J_2$.

To estimate the error of the numerical solution, we consider the discrepancy of the boundary condition at points between the collocation points, which is calculated in the linear metric as

$$\Delta\left(\tilde{\theta}_m, \varphi\right) = \left|A\left(\theta_n, \tilde{\theta}_m, \varphi\right) \cdot I - U^0\left(\tilde{\theta}_m, \varphi\right)\right|, \qquad (3.22)$$

where A is the SLAE matrix and $\tilde{\theta}_m = \frac{(m-1)\pi}{N}$, $n, m = \overline{1, N}$. The rate of the discrepancy decrease with increasing approximation level N allows us to estimate the internal convergence of the computational algorithm. Everywhere below in the calculations, the discrepancies are constructed in the plane $\varphi = \text{const}$, where they take maximal values.

3.3.1 Illustration of the Necessity to Consider the Singularities of the Wave Field Analytic Continuation in NFM

Now we use several examples of solving the scalar and vector problems of diffraction at bodies of revolution to illustrate the assertion that the surface Σ must surround the set $\bar{A}$ of singularities of the analytic continuation of the field scattered at the body bounded by the surface S and that it is also necessary to construct this surface by an analytic deformation in NFM.

In these examples, the surface Σ is constructed by three different methods: (a) by an analytic deformation of S [26, 60, 72] (also see Chapter 2) such that it surrounds the set $\bar{A}$ of the wave field singularities, and also by shifting S by the same value Δ for all values of the angle θ such that it surrounds (b) and it does not surround (c) the set $\bar{A}$. Integral equations (3.5) and (3.20) were solved numerically by using the algorithm discussed in Section 3.2.1 with transition to discrete sources.

We consider the problem of diffraction of the plane wave

$$U^0(\theta, \varphi) = \exp\{-ikr[\sin\theta\sin\theta_0\cos(\varphi - \varphi_0) + \cos\theta\cos\theta_0]\} \qquad (3.23)$$

propagating in the direction determined by the angles θ_0, φ_0. The coefficients of the Fourier expansion of the plane wave (3.23) have the form

$$U_0^s(\theta) = (-i)^s \exp(-ik\rho(\theta)\cos\theta\cos\theta_0 - is\varphi_0) \cdot J_s(k\rho(\theta)\sin\theta\sin\theta_0).$$

Figure 3.1 illustrates the results of calculations for an acoustically soft spheroid with semiaxes $ka = 1$, $kc = 5$, $Z = 0$ for the incidence angles $\theta_0 = \pi/2$, $\varphi_0 = 0$, $N = N_1 = 128$, where N is the level of approximation of the unknown function, and N_1 is the number of nodes used to calculate the inverse Fourier transform of Green's function $G_0\left(\vec{r}, \vec{r}'\right)$.

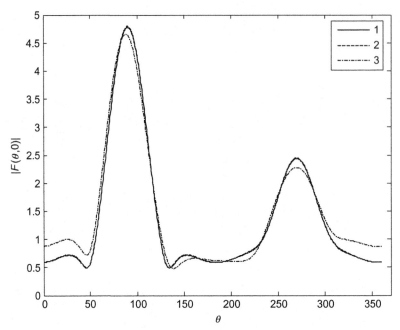

Figure 3.1 Scattering pattern for acoustically soft spheroid with semiaxes $ka = 1, kc = 5$.

Here we introduce the following notation: 1 denotes the surface S_δ obtained by an analytic deformation of the scatterer boundary, 2 denotes the surface S_δ obtained by a nonanalytic deformation of S such that it surrounds the singularities (the distance $k\Delta$ from the surface S_δ to the set of singularities is equal to 10^{-3}), and 3 denotes the surface S_δ obtained by a nonanalytic deformation of S such that it does not surrounds the singularities (the distance $k\Delta$ from the surface S_δ to the set of singularities is equal to -0.1).

We clearly see that in the case where the singularities are not surrounded, the result is false, i.e., pattern 3 even has no symmetry that must appear for $\theta_0 = \pi/2$. Thus, the obtained results confirm the assertion that to obtain correct results, the surface S_δ must necessarily surround the singularities of the analytic continuation of the wave field.

Figure 3.2 illustrates the results of calculating the scattering pattern for the same problem in the case of impedance spheroid with $Z = 1000\zeta i$. Here we do not present the results obtained in the case where the surface Σ does not surround the set of singularities, because even if this set is surrounded (in the case of nonanalytic deformation), the result is still false. The true result in the case of nonanalytic deformation of the boundary that surrounds the

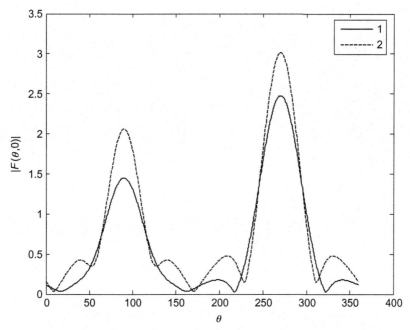

Figure 3.2 Scattering pattern for impedance spheroid with $Z = 1000\zeta i$.

singularities is obtained only for $N = N_1 = 512$. But in the case where the surface Σ does not surround the set $\bar{A}$ of singularities, the computational algorithm does not give correct results for any value of N.

Figures 3.3–3.5 illustrate the results of calculations of the discrepancy when the surface Σ is constructed by methods (a)–(c), respectively, for the considered problem in the case of an acoustically soft spheroid for different values of N (64, 128, 256, 512).

We see that the analytic deformation ensures a high rate of convergence of the numerical algorithm (because the discrepancies noticeably decrease with increasing N), while in the case of nonanalytic deformation, the convergence is significantly worse. Moreover, the discrepancies obtained by analytic deformation have practically no overshoots, while the discrepancies obtained by nonanalytic deformation have overshoots in a wide range of θ and do not practically decrease as N increases. We can also see that the discrepancies in Figs. 3.4 and 3.5 are very similar to each other, have the same order of magnitude, and, in the case where the singularities are not surrounded, the pattern is false. This means that the pattern irregularity is not related to the deficiency at the collocation points as in the case of pattern 2 in Fig. 3.2 but is related to the false choice of the surface Σ.

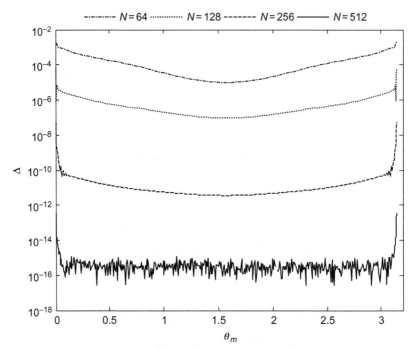

Figure 3.3 Discrepancy for the surface Σ constructed by method (a).

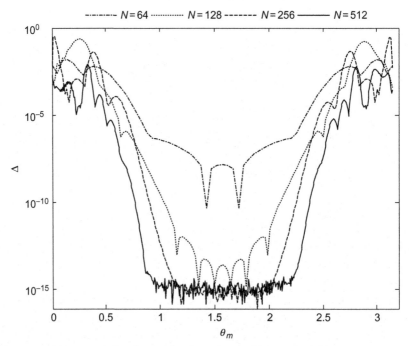

Figure 3.4 Discrepancy for the surface Σ constructed by method (b).

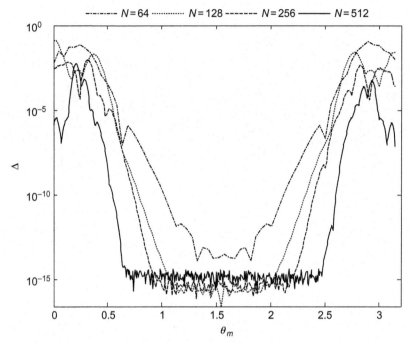

Figure 3.5 Discrepancy for the surface Σ constructed by method (c).

Now we verify to what extent the null field condition is satisfied on the surfaces other than Σ in the case where the impedance on the scatterer boundary is nonzero. Figure 3.6 presents the results of calculating the discrepancy on different surfaces lying inside the considered spheroid with semiaxes $ka = 1, kc = 5$ in the case where the surface Σ was constructed by an analytic deformation S. The insert in this figure presents the value of the parameter δ of the analytic deformation, the lowest curve corresponds to the discrepancy on the surface Σ calculated at points between the collocation points, and the uppermost curve corresponds to the discrepancy on the surface nearly adjacent to S. One can see that the null field condition is satisfied everywhere with an appropriate accuracy, despite the fact that the total field is nonzero on the surface S itself. Near the surface S, the fulfillment of the null field condition is worse by several orders than on the surface Σ, but in Section 3.3.2, it is shown that the value of discrepancy on Σ reflects the accuracy of calculations.

Figure 3.7 illustrates the results of similar calculations in the case where the surface Σ is constructed by a nonanalytic deformation of S and surrounds the singularities for $N = N_1 = 512$. In this figure, the values $k\Delta r$ of the

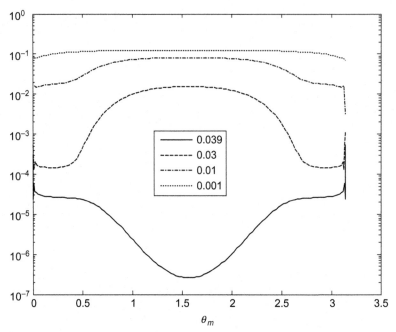

Figure 3.6 Discrepancy on different surfaces inside spheroid. The surface Σ is constructed by an analytic deformation of S.

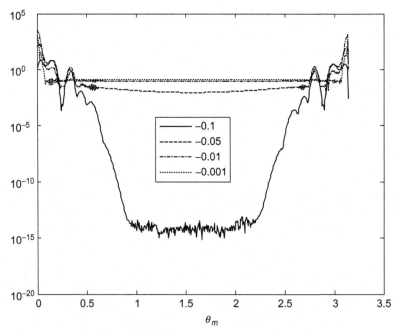

Figure 3.7 Discrepancy on surfaces inside spheroid. The surface Σ is constructed by a nonanalytic deformation of S and surrounds the singularities.

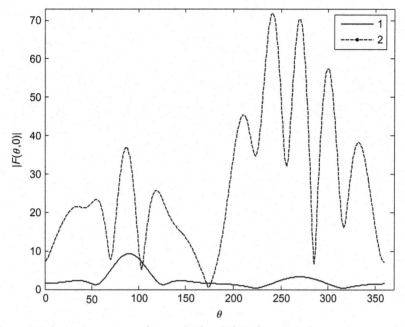

Figure 3.8 Scattering patterns for two-leaf multifoil of revolution.

distance from the corresponding surface to the scatterer boundary S are indicated near the curves. One can see that the discrepancy is acceptable in the whole but is attained only as the number of discrete sources is four times greater.

Figure 3.8 presents the scattering patterns for a two-leaf multifoil of revolution, i.e., a body formed by revolution of the curve $\rho(\theta) = b + a\cos 2\theta$, with the parameters $ka = 2$, $kb = 3$; the patterns were obtained for $N = N_1 = 256$ and $Z = 0$. Here, curve 1 was calculated by constructing the surface Σ by an analytic deformation till the singularities, and curve 2 was obtained by constructing the surface Σ as a sphere of radius 0.999, just as, for example, in the T-matrix method, where the singularities, which lie at the distance 3.62 from the two-leaf multifoil center in our case, are not surrounded.

In the studies where the applicability conditions of the NFM are discussed, it is noted that this method cannot be applied to strongly prolonged scatterers. In particular, it has been noted [29] that the standard NFM allows one to calculate the particle scattering characteristics if the ratio of the maximal to the minimal size does not exceed 4:1.

Figures 3.9 and 3.10 illustrate the scattering patterns and discrepancies obtained by solving the problem of diffraction of a plane wave incident at

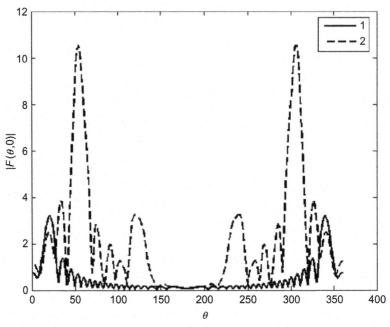

Figure 3.9 Scattering patterns for acoustically hard spheroid with semiaxes $ka = 3.004$, $kc = 30.15$.

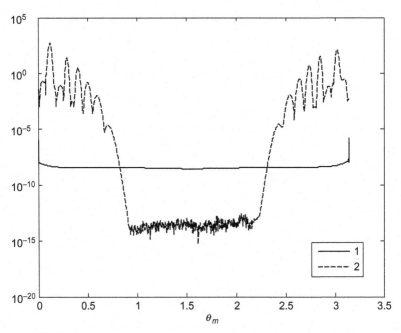

Figure 3.10 Discrepancies.

the angles $\varphi_0 = 0$, and $\theta_0 = 0$ at an acoustically hard (the Neumann condition, i.e., $|Z| \gg 1$ in (3.2)) spheroid with semiaxes $ka = 3.004$ and $kc = 30.15$ for $N = N_1 = 512$. Curves 1 correspond to the construction of the surface Σ by analytic deformation of the scatterer boundary, and curves 2 correspond to the surface Σ obtained by a nonanalytic deformation of S such that the singularities are surrounded. We do not present the results obtained without surrounding the set of singularities by the surface Σ, because even if this set were surrounded, the result would still be false. In the case of nonanalytic deformation of the boundary with surrounded singularities, the true result can be obtained only for $N = N_1 = 1024$. Thus, we conclude that the correct realization of the NFM (MNFM), where the surface Σ surrounds the set of singularities, also permits solving the problems of diffraction at strongly prolonged scatterers.

Now we pass to the vector case. We consider the problem of diffraction of a plane electromagnetic wave (whose electric vector lies in the plane of incidence) incident at the angles $\varphi_0 = 0$ and $\theta_0 = \pi/2$ at an ideally conducting (i.e., $Z = 0$ in (3.7)) spheroid with semiaxes $ka = 1$, $kc = 5$. Figure 3.11 illustrates the results of calculating the scattering patterns that were obtained by constructing the surface Σ by an analytic deformation of the surface S.

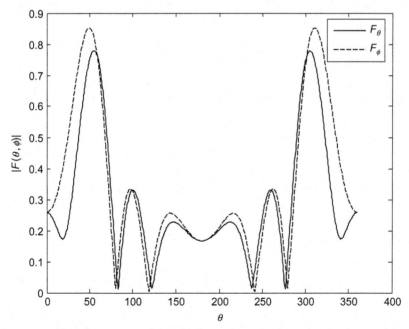

Figure 3.11 Scattering pattern for perfectly conducting spheroid with semiaxes $ka = 1$, $kc = 5$. The surface Σ is constructed by an analytic deformation of S.

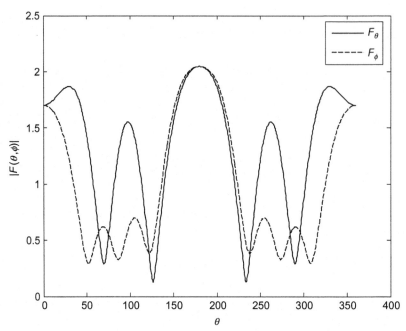

Figure 3.12 Scattering pattern for perfectly conducting spheroid with semiaxes $ka = 1$, $kc = 5$. The surface Σ is constructed by a nonanalytic deformation of S and surrounds the singularities.

The calculations were performed for $N = N_1 = 256$. In Fig. 3.12, we present the results of calculating the scattering pattern obtained for $N = N_1 = 256$ by a nonanalytic deformation of the boundary with surrounded singularities. One can see that the results are absolutely false. In this case, the correct results can be obtained only for $N = N_1 = 1024$. Figure 3.13 illustrates the solution discrepancies calculated in the case where the surface Σ was constructed by an analytic deformation of the surface S to the wave field singularities. Here, the lowest curve corresponds to the discrepancy on the surface Σ calculated at the points lying between the collocation points, and the uppermost curve corresponds to the surface almost adjacent to S. We can see that the null field condition is also satisfied with an appropriate accuracy in the vector case also.

In the case where the surface Σ does not surround the set of the wave field singularities, the true results cannot be obtained for any N.

3.3.2 Null Field Method and the Method of Auxiliary Currents

A criterion for the correctness (accuracy) of the problem solution in the MAC is usually the value of the boundary condition discrepancy $\Delta(\theta)$ at

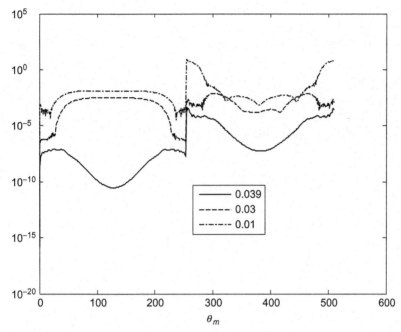

Figure 3.13 Solution discrepancies for the surface Σ constructed by an analytic deformation of S till the wave field singularities.

the scatterer boundary points in the middle between the collocation points. In NFM, the situation is somewhat different, because in this method, the null field condition rather than the boundary condition must be satisfied on the surface Σ, where we choose the observation points in integral equation (3.3) and hence the collocation points when passing to the system of algebraic equations.

For the same diffraction problem and the same size N of the system of algebraic equations, the boundary condition discrepancy on the surface S in MAC is worse than the null field condition discrepancy on the surface Σ in NFM; one of causes is that the surface Σ is smaller in this case. The farther is the surface Σ from the surface S, the greater is the difference in the discrepancies. We note that when the system size becomes large so that the discrepancy is of the order of the computer accuracy (i.e., the greatest possible accuracy is attained), the NFM discrepancy becomes close to the MAC discrepancy but it is still always worse than the MAC discrepancy. (All this is confirmed by the calculations given below.)

But if we calculate the null field discrepancy in NFM on the surface S, then it is obviously much worse than the MAC discrepancy. This allows us to conclude that the NFM accuracy is lower than the MAC accuracy. However, as follows from the numerical results given below, this is an erroneous conclusion, and precisely the null field condition discrepancy on the surface Σ (if the choice of this surface is correct) is the best characteristic of the accuracy in the NFM calculations.

Let us consider the problem of diffraction of the plane wave (3.23) incident at the angles $\theta_0 = 0$, $\varphi_0 = 0$ at a spheroid, at a superellipsoid with the axial cross-section of the form $\frac{x^{2m}}{a^{2m}} + \frac{z^{2m}}{c^{2m}} = 1$ for $m = 8$, and at a 10-leaf multifoil of revolution whose axial cross-section has the form $\rho(\theta) = c + a \cdot \cos 10\theta$ for $Z = 0$. All calculations are performed by using the technique of discrete sources, both in the case of MAC and in the case of NFM. The surface Σ is constructed by an analytic deformation of S.

Figure 3.14 illustrates the discrepancies obtained for $N = 50$ (from now on, we assume that $N = N_1$) for the spheroid with semiaxes $ka = 4$, $kc = 10$: curve 1 corresponds to the discrepancy on S in MAC, and curve 2 corresponds to the discrepancy on Σ in NFM. One can see that the discrepancy on Σ in NFM is approximately by an order of magnitude less than the discrepancy on S in MAC. In this case, the analytic continuation singularities are at a rather far distance from the surface S, and hence the discrepancy on Σ is better. We perform similar calculations with $N = 50$ for a spheroid with semiaxes $ka = 2$, $kc = 10$ in the case where the singularities of the analytic

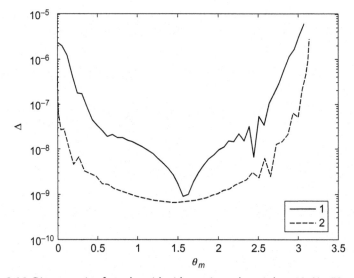

Figure 3.14 Discrepancies for spheroid with semiaxes $ka=4$, $kc=10$, $N=50$.

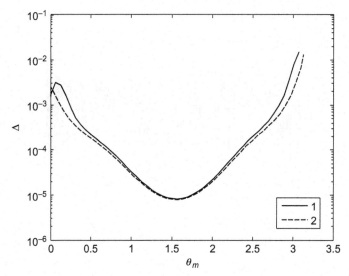

Figure 3.15 Discrepancies for spheroid with semiaxes $ka=2$, $kc=10$, $N=50$.

continuation are located much closer to S. The results of these calculations are shown in Fig. 3.15, where the notation is preserved. One can see that since the surfaces S and Σ are close to each other, the discrepancies on them are also close to each other.

Figure 3.16 illustrates the results of similar calculations for a spheroid with semiaxes $ka=4$, $kc=10$ for $N=200$. One can see that the MAC and NFM discrepancies become closer to each other, but the NFM discrepancy is still somewhat better. Figure 3.17 shows the null field condition discrepancy in NFM on the surface S for the same problem for $N=50$ (curve 1) and for $N=200$ (curve 2). One can see that it is by many orders of magnitude greater than the corresponding discrepancy on Σ (curve 2 in Fig. 3.16).

As previously noted, the NFM discrepancy on Σ is more demonstrative and better reflects the accuracy of the obtained solution. To verify this, we compare the values of the patterns obtained by MAC ($f_1(\theta)$) and by NFM ($f_2(\theta)$) at each point. For this, we calculate

$$\Delta f(\theta) = |f_1(\theta) - f_2(\theta)|,$$

and also verify the optical theorem [8]

$$\frac{k^2 \sigma_S}{4\pi} \equiv \frac{1}{4\pi} \int_\Omega |f(\theta, \varphi; \theta_0, \varphi_0)|^2 \sin\theta \, d\theta \, d\varphi = -\mathrm{Im} f(\theta = \theta_0, \varphi = \varphi_0), \quad (3.24)$$

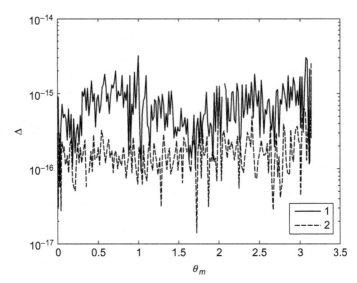

Figure 3.16 Discrepancies for spheroid with semiaxes $ka = 4$, $kc = 10$, $N = 200$.

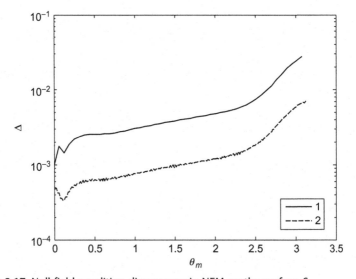

Figure 3.17 Null field condition discrepancy in NFM on the surface S.

where Ω is the surface of the unit sphere. The quantity $\frac{k^2\sigma_S}{4\pi}$ is the integral scattering cross-section. To estimate the accuracy of the optical theorem, we calculate the quantity

$$\Delta_1 = \left| \frac{k^2\sigma_S}{4\pi} + \mathrm{Im}f(\theta=\theta_0, \varphi=\varphi_0) \right|. \tag{3.25}$$

Figures 3.18 and 3.19 present the graphs of $\Delta f(\theta)$ obtained by solving the problem of diffraction at a spheroid with semiaxes $ka = 4$, $kc = 10$ for $N = 50$ and $N = 200$, respectively.

Figures 3.20 and 3.21 present the results of similar calculations for the superellipsoid with the same semiaxes for $N = 200$ and $N = 500$, respectively. Figures 3.22 and 3.23 present the results of similar calculations for a 10-leaf multifoil of revolution with the parameters $ka = 3$, $kc = 7$ obtained for $N = 200$ and $N = 500$.

One can see that for the limit values of N (i.e., for the values at which the discrepancy is of the order of the computer accuracy), the patterns obtained by both methods coincide with the computer accuracy. For example, for $N = 50$ in the case of a spheroid, the difference in the patterns is equal to 10^{-4}, and then the discrepancy on Σ in NFM differs from the discrepancy on S in MAC approximately by an order (see Fig. 3.5), while the discrepancy on S in NFM differs from the corresponding discrepancy

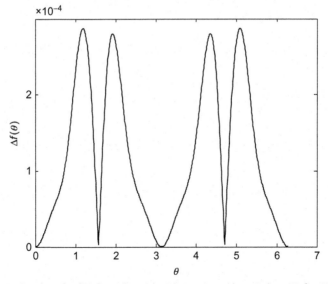

Figure 3.18 Graphs of $\Delta f(\theta)$ for spheroid with semiaxes $ka = 4$, $kc = 10$ for $N = 50$.

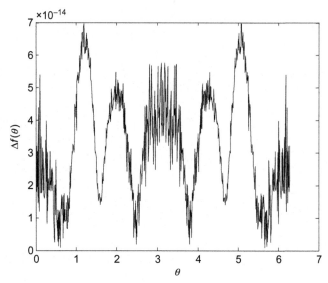

Figure 3.19 Graphs of $\Delta f(\theta)$ for spheroid with semiaxes $ka = 4$, $kc = 10$ for $N = 200$.

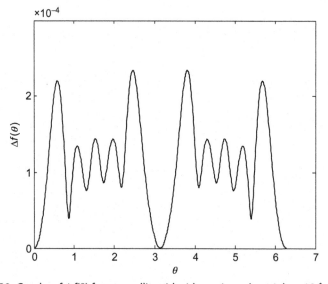

Figure 3.20 Graphs of $\Delta f(\theta)$ for superellipsoid with semiaxes $ka = 4$, $kc = 10$ for $N = 200$.

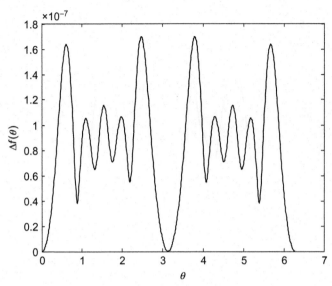

Figure 3.21 Graphs of $\Delta f(\theta)$ for superellipsoid with semiaxes $ka = 4$, $kc = 10$ for $N = 500$.

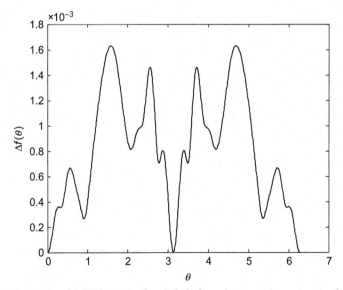

Figure 3.22 Graphs of $\Delta f(\theta)$ for 10-leaf multifoil of revolution with parameters for $ka = 3$, $kc = 7$ for $N = 200$.

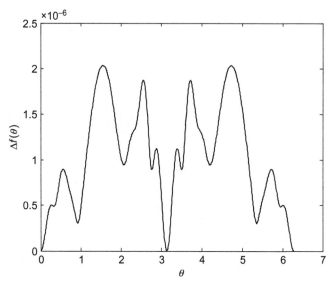

Figure 3.23 Graphs of $\Delta f(\theta)$ for 10-leaf multifoil of revolution with parameters $ka = 3$, $kc = 7$ for $N = 500$.

in MAC approximately by three or four orders. This means that precisely the discrepancy on the surface Σ estimates the accuracy in NFM. It also follows from Figs. 3.18–3.23 that the accuracy in MAC and in NFM is approximately the same. But for a small N, there is still a difference in these patterns, and although this difference is insignificant, we still verify the optical theorem to estimate the accuracy of the pattern calculations by both methods more precisely.

In Table 3.1, we present the results of this verification of the optical theorem (the value of Δ_1) for the considered geometries and for different N.

One can see that even, for example, for $N = 50$ in the case of a spheroid, the values of Δ_1 obtained in MAC and in NFM differ only in the seventh

Table 3.1 Verification of the optical theorem

		MAC	NFM
Spheroid	$N = 50$	$1.4921909112786 \times 10^{-5}$	$1.4955288836670 \times 10^{-5}$
	$N = 200$	$9.7699626167014 \times 10^{-14}$	$7.8159700933611 \times 10^{-14}$
Superellipsoid	$N = 200$	$6.1738351870844 \times 10^{-5}$	$6.1756244143396 \times 10^{-5}$
	$N = 500$	$4.7419781878943 \times 10^{-8}$	$4.7419828064221 \times 10^{-8}$
10-Leaf multifoil	$N = 200$	0.0035550797802	0.0035539490393
	$N = 500$	$4.5412489910746 \times 10^{-6}$	$4.5412739879680 \times 10^{-6}$

digit, and for $N = 200$ in the case of a 10-leaf multifoil, the values of $\varDelta_1$ differ in the sixth digit. This fact again confirms that the accuracy in MAC and in NFM is approximately the same.

We note that, as calculations showed [22, 78–80], the null field condition can be satisfied sufficiently well (in the root-mean-square metric) even in the case of an incorrectly obtained solution. Therefore, in NFM it is particularly important to satisfy the requirement that the set of the wave field singularities must be surrounded by the surface $\varSigma$. As an additional criterion for the correctness of the obtained results, we can consider the verification of the null field condition on the surfaces between $\varSigma$ and S [79].

It should be noted that precisely a surface, i.e., a set of the same dimension as S, must be chosen as the surface $\varSigma$ in both MAC and NFM. This remark is not quite trivial. The point is that, according to some recommendations (see [22, 78–80]), to obtain the most fast-operating and stable algorithm, it is expedient to construct the surface $\varSigma$ by an analytic deformation of the scatterer boundary S so that it maximally closely "surrounds" the set $\bar{A}$ of the wave field singularities, just as in MAC [26]. But in the case where this set consists of a single point (e.g., as in the case of a sphere) or lies completely inside a sphere of an extremely small radius (see below), the above recommendation must be refined. In this case, when solving the problem numerically, we arrive at the following computational contradiction. On the one hand, the surface $\varSigma$ is very small and hence the number of collocation points on it must be small, because otherwise, the algorithm "interprets" this surface as a single point in the sense that the matrix elements corresponding to different points on $\varSigma$ are close to each other in value. On the other hand, it is impossible to deal with few collocation points, because the corresponding number of sources on S is also small, which leads to a rough approximation of the field distribution on S. Naturally, this destroys the algorithm and, for example, in the case of a sphere, if we take a sphere of a very small radius as the surface $\varSigma$, we obtain false results. The minimally admissible value of the radius (we denote it by $r_{\min}$) of the sphere $\varSigma$ depends on the radius of the surface S, i.e., the greater this radius, the greater the value of $r_{\min}$, which can be determined by numerical experiments. For example, for a sphere of radius $ka = 4\pi$, we have $kr_{\min} \approx 0.3$, and for a sphere of radius $ka = \pi$, we have $kr_{\min} \approx 0.1$. However, this contradiction can be removed by using algorithms of extra accuracy.

Figure 3.24 shows the scattering patterns for sphere (1), oblate spheroid (2), and trefoil of revolution (3) radiated by the plane wave (3.23) incident at the angles $\theta_0 = 0$, $\varphi_0 = 0$. The sphere radius is $ka = 4\pi$, the spheroid

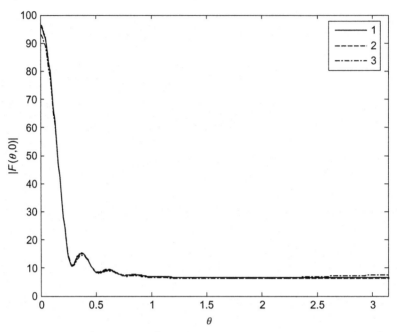

Figure 3.24 Scattering patterns for sphere (1), oblate spheroid (2), and trefoil of revolution (3).

dimensions are $ka = 4\pi$, $kc = 4\pi - 0.5$, the axial cross-section of the trefoil has the form $\rho(\theta) = c + a \cdot \cos 3\theta$, and its dimensions are $ka = 0.25$, $kc = 4\pi - 0.25$. The calculations were performed by the NFM for the number of discrete sources $N = 200$. The discrepancy on the surface Σ is of the order of 10^{-15}.

One can see that the scattering patterns practically coincide for the sphere and spheroid, and the trefoil pattern differs from them only insignificantly. In the calculations, the surface Σ for the spheroid and trefoil was constructed by an analytic deformation until the singularities were reached, and, according to the above, a sphere of radius 0.3 was taken as the surface Σ. The maximal distance of points of the surface Σ is then equal to $kr_0 = kf \cong 0.28$ (where f is the interfocus distance) for the spheroid and to $kr_0 \cong 0.12$ for the trefoil. One can see that this distance approximately corresponds to the minimally admissible value of the radius of the sphere Σ. Moreover, it follows from the results shown in Fig. 3.24 that, even for the figures that slightly differ from the sphere and whose scattering patterns practically coincide with the scattering pattern of the sphere, the algorithm based on the construction of the surface

Σ by an analytic deformation of the scatterer boundary with the maximally close surrounding of the set of singularities is sufficiently reliable and ensures correct results.

3.4 *T*-MATRIX METHOD

3.4.1 Derivation of Basic Relations

Now we present a correct of the *T*-matrix method. As previously noted, TMM is a modification of NFM, where the null field condition is assumed to be posed on a sphere which lies completely inside the scatterer.

For simplicity, we assume that the Dirichlet condition $U|_S = 0$ is satisfied on the boundary S. Equation (3.5) then becomes

$$U^0(\vec{r}) - \int_S \frac{\partial U(\vec{r}')}{\partial n'} G_0(\vec{r}, \vec{r}') \, ds' = 0, \quad M(\vec{r}) \in \Sigma. \tag{3.26}$$

We introduce the notation

$$\frac{\partial U}{\partial n'}\bigg|_S = \frac{1}{\kappa \rho} \frac{4\pi i}{k} I(\theta', \varphi') \sin \theta', \quad ds' = \kappa \rho d\theta' d\varphi',$$

$$\kappa = \sqrt{\left(\rho^2 + \rho_{\theta'}'^2\right) \sin^2\theta' + \rho_{\varphi'}'^2},$$

where $r = \rho(\theta, \varphi)$ is the equation of boundary S. With this notation taken into account, the integral equation of the null field in the case under study has the form

$$\frac{4\pi i}{k} \int_0^{2\pi} I(\theta', \varphi') G_0(\vec{r}, \vec{r}') \sin \theta' d\theta' d\varphi'\bigg|_\Sigma = U^0(\vec{r})\big|_\Sigma, \tag{3.27}$$

where Σ is a simple closed surface inside S.

As previously noted, in TMM, a sphere of radius $r = r_0$ completely lying inside S is chosen as Σ [5, 8, 19].

Let $U^0(\vec{r})$ be the field of the plane wave (3.23) propagating at the angles θ_0, φ_0 with the axes Oz and Ox, respectively,

$$U^0(\vec{r}) = \sum_{n=0}^{\infty} \sum_{m=-n}^{n} (-i)^n N_{nm} j_n(kr) P_n^m(\cos\theta) P_n^m(\cos\theta_0) e^{im(\varphi - \varphi_0)}, \tag{3.28}$$

where $N_{nm} = \frac{2}{2n+1} \frac{(n+m)!}{(n-m)!}$.

We set

$$
I(\theta', \varphi') = \sum_{\nu=0}^{\infty} \sum_{\mu=-\nu}^{\nu} b_{\nu\mu} P_{\nu}^{\mu}(\cos\theta') e^{i\mu\varphi'},
$$

then from Equation (3.8), taking into account the fact that Σ is a sphere inside S, we obtain

$$
\sum_{\nu=0}^{\infty} \sum_{\mu=-\nu}^{\nu} H_{nm,\nu\mu} b_{\nu\mu} = a_{nm}, \quad n = 0, 1, \ldots; m = 0, \pm 1, \ldots, \pm n \quad \text{or} \quad H\bar{b} = \bar{a},
$$

$$
(3.29)
$$

where

$$
H_{nm,\nu\mu} = \int_0^{2\pi} \int_0^{\pi} h_n^{(2)}(k\rho(\theta, \varphi)) P_n^m(\cos\theta) P_{\nu}^{\mu}(\cos\theta) e^{i(\mu-m)\varphi} \sin\theta d\theta d\varphi,
$$

$$
a_{nm} = -(-i)^n P_n^m(\cos\theta_0) e^{-im\varphi_0}.
$$

Everywhere in the domain $r > r_1 \equiv \max\limits_{\theta,\varphi} \rho(\theta, \varphi)$, we have the expansion

$$
U^1(\vec{r}) = -\frac{4\pi i}{k} \int_0^{2\pi} \int_0^{\pi} I(\theta', \varphi') G_0(\vec{r}, \vec{r}') d\theta' d\varphi'
$$

$$
= \sum_{n=0}^{\infty} \sum_{m=-n}^{n} c_{nm} h_n^{(2)}(kr) P_n^m(\cos\theta) e^{im\varphi}, \qquad (3.30)
$$

where

$$
c_{nm} = \sum_{\nu=0}^{\infty} \sum_{\mu=-\nu}^{\nu} G_{nm,\nu\mu} b_{\nu\mu}, \qquad (3.31)
$$

and

$$
G_{nm,\nu\mu} = -N_{nm} \int_0^{2\pi} \int_0^{\pi} j_n(k\rho(\theta, \varphi)) P_n^m(\cos\theta) P_{\nu}^{\mu}(\cos\theta) e^{i(\mu-m)\varphi} \sin\theta d\theta d\varphi.
$$

As a result, relations (3.30) and (3.31) imply

$$
\bar{c} = G\bar{b} = GH^{-1}\bar{a} \equiv T\bar{a}.
$$

Thus, we expressed the coefficients of expansion (3.30) of the scattered field in terms of the coefficients $\bar{a}$ of the incident field. The matrix $T = GH^{-1}$ is therefore called a T-matrix, and the described technique for solving the diffraction boundary-value problem is called the *T*-matrix method.

For the scattering pattern $g(\theta, \varphi)$, we have

$$g(\theta, \varphi) = \sum_{n=0}^{\infty} \sum_{m=-n}^{n} c_{nm} i^n P_n^m(\cos\theta) e^{im\varphi}. \tag{3.32}$$

In the case of a body of revolution, i.e., for $\rho(\theta, \varphi) = \rho(\theta)$, system (3.29) and relation (3.31) can be simplified as

$$\sum_{\nu=|m|}^{\infty} H_{nm,\nu m} b_{\nu m} = a_{nm}, \quad n = 0, 1, \ldots; \; m = 0, \pm 1, \ldots, \pm n, \tag{3.29a}$$

$$c_{nm} = \sum_{\nu=|m|}^{\infty} G_{nm,\nu m} b_{\nu m}. \tag{3.31a}$$

Doicu *et al.* [28] noted that as the ratio of the maximal to the minimal scatterer size increases, the rate of convergence and the TMM accuracy become significantly worse because of the ill-posedness of the matrices. Similarly, Shenderov [8] noted that the standard TMM is bounded by the value of this ratio in the order of four. However, the TMM applicability is determined not by the ratio of the maximal to the minimal scatterer size as was noted above, but by the fact that the surface Σ (a sphere or a circle in the two-dimensional case in TMM) must surround the set of singularities $\bar{A}$ of the analytic continuation of the wave field. This is clear because expansion (3.30) is used here implicitly to solve integral equation (3.27), and this expansion exists only in the domain $r > R_0$, where R_0 is the distance from the point of the set $\bar{A}$, which is most remote from the origin [9, 10] (see also Chapter 1).

Thus, the above-described TMM algorithm is correct only if

$$R_0 < \min_{\theta, \varphi} \rho(\theta, \varphi). \tag{3.33}$$

Inequality (3.33) is satisfied in the case of Rayleigh scatterers [9, 10].

3.4.2 Numerical Studies

Now we illustrate the importance of restriction (3.33). First, we consider the problem of plane wave (3.23) diffraction at a spheroid with semiaxes $ka = 2$ and $kc = 4$ with the Dirichlet boundary condition satisfied on its boundary. Figure 3.25 shows the scattering patterns for the considered problem for different N: curve 1 corresponds to $N = 30$, curve 2 corresponds to $N = 40$,

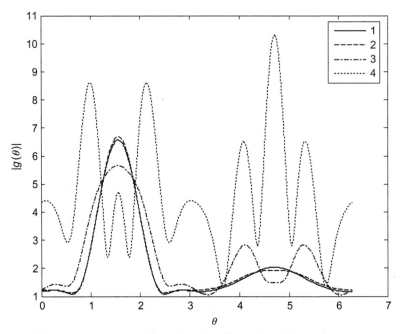

Figure 3.25 Scattering patterns for spheroid with semiaxes $ka = 2$, $kc = 4$.

curve 3 corresponds to $N = 41$, and curve 4 corresponds to $N = 42$. The pattern for $N = 30$ graphically coincides with the pattern obtained by NFM (with the singularities taken into account) for $N = 100$ (N in NFM is taken larger to ensure a high accuracy of the result). Thus, for a spheroid of the considered dimensions, TMM principally permits obtaining results with a good accuracy, because, in this case, only a small part of the set of singularities $\bar{A}$, which is the interfocus segment, lies beyond the inscribed sphere. However, the more elongated (oblate) the spheroid is, the greater the part of the set of singularities beyond the inscribed sphere is and the greater the TMM error is.

Despite the fact that TMM in this case allows one to obtain a good accuracy, it follows from Fig. 3.25 that the pattern for $N = 40$ has a visual error that significantly increases for $N = 41$ and the algorithm is completely destroyed already for $N = 42$. This result confirms the importance of inequality (3.33) in TMM. The spheroid in question is not a Rayleigh spheroid and inequality (3.33) cannot be satisfied in this case. This implies that the considered expansions diverge, which in turn is reflected in the pattern for $N > 40$. We again stress that the TMM accuracy is influenced not by the ratio of the maximal to minimal scatterer size but by the character of location of the singularities (which, in the case of spheroids, is directly related to the ratio of its dimensions). If the scatterer satisfies the Rayleigh hypothesis, then TMM allows one to obtain the results with a prescribed accuracy; otherwise, the TMM error depends on the part of the set of singularities outside the inscribed sphere, i.e., TMM does not guarantee any acceptable accuracy of the obtained results.

Now we consider the problem of the plane wave (3.23) diffraction at the 10-leaf multifoil with the parameters $ka = 1$, $kc = 4$. Figure 3.26 presents the scattering patterns for the considered problem obtained by TMM for $N = 31$ (curve 1) and by NFM (with the singularities taken into account) for $N = 100$ (curve 2).

One can see that the pattern calculated by TMM has a noticeable error, and this error increases with decreasing or increasing N. Namely, although TMM for some $N(\sim 30 \div 37)$ permits us to obtain a pattern close to the correct one, it still seems impossible to attain a high accuracy of the results. This can be explained by the fact that although the ratio of the maximal to minimal size (it is equal to 5:3) is here less than in the preceding example, the

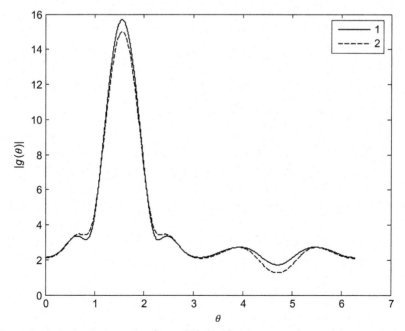

Figure 3.26 Scattering patterns for multifoil with parameters $ka = 1$, $kc = 4$.

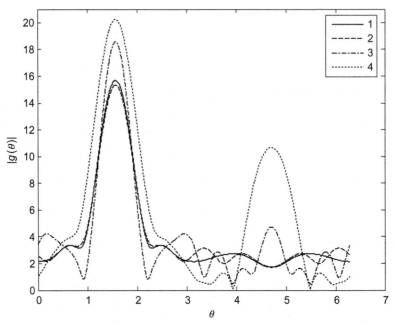

Figure 3.27 Scattering patterns for multifoil with parameters $ka = 1$, $kc = 4$ obtained by TMM.

considered 10-leaf multifoil is not a Rayleigh scatterer, and a considerable part of the set of singularities is not surrounded by the inscribed sphere.

In Fig. 3.27, we present the scattering patterns obtained by TMM for the this problem for different N: curve 1 corresponds to $N = 31$, curve 2 corresponds to $N = 37$, curve 3 corresponds to $N = 38$, and curve 4 corresponds to $N = 40$.

One can see that the pattern is more or less correct for $N = 31$, although with a noticeable error, but already for $N \geq 37$, the algorithm is completely destroyed.

Thus, we can conclude that the numerical realization of the NFM is correct only under the condition that the surface on which the null-field condition is posed surrounds the set of singularities of the wave field analytic continuation. An optimal method for constructing such a surface is the analytic deformation of the scatterer boundary. Moreover, we can state that the T-matrix method [5, 8, 19, 20] is correct only for Rayleigh scatterers.

3.4.3 Modified *T*-Matrix Method

Now we consider another approach based on the MNFM, which was briefly discussed in Section 3.1 (also see [22, 78–82]).

For simplicity, we consider the two-dimensional case. The integral equation (3.5) becomes

$$\int_0^{2\pi} J(\varphi')\left\{H_0^{(2)}\left(k|\vec{r}-\vec{r}'|\right) - \frac{W}{k}\frac{\partial H_0^{(2)}\left(k|\vec{r}-\vec{r}'|\right)}{\partial n'}\right\}d\varphi' = -u^0(\vec{r}),\ \vec{r}\in\Sigma,$$

(3.34)

where $Z = i\zeta W$, $k = \omega\sqrt{\varepsilon_0\mu_0}$ is the wave number, ζ is the wave resistance (impedance) in the environment, and Σ is a simple closed curve inside the scatterer boundary S.

According to MNFM, the contour Σ in integral equation (3.34) must surround the set $\bar{A}$.

Now we solve Equation (3.34) by the method of discrete sources [22]. According to this method, we replace the integral in the left-hand side of Equation (3.34) by the sum of point sources localized on S,

$$\int_0^{2\pi} J(\varphi')\left\{H_0^{(2)}\left(k|\vec{r}-\vec{r}'|\right) - \frac{W}{k}\frac{\partial H_0^{(2)}\left(k|\vec{r}-\vec{r}'|\right)}{\partial n'}\right\}d\varphi' \cong \sum_{n=1}^{N} a_n K(\vec{r};\vec{r}_n),$$

(3.35)

and then equate the left- and right-hand sides of Equation (3.34) at the collocation points on Σ. As a result, we obtain an algebraic system of the form

$$\sum_{n=1}^{N} a_n K(\vec{r}_m;\vec{r}_n) = -u^0(\vec{r}_m),$$

(3.36)

where

$$\vec{r}_n = \{r_S(\varphi_n),\varphi_n\} = \{r_{Sn},\varphi_n\},\ \vec{r}_m = \{r_\Sigma(\varphi_m),\varphi_m\} = \{r_{\Sigma m},\varphi_m\}.$$

Here $r_S(\varphi)$ is the equation of the scatterer boundary S and $r_\Sigma(\varphi)$ is the equation of the curve Σ in polar coordinates. According to MNFM [22, 23, 78, 79], the curve Σ is constructed (just as in the modified MAC [26]) by an analytic deformation of the scatterer boundary S to the wave field singularities. The algorithm for such a deformation is described in detail in [22, 23, 26, 78–80].

According to Equation (3.34), everywhere outside S, the diffraction field in the case under study is represented as the sum of point sources

$$U^1(\vec{r}) \cong \sum_{n=1}^{N} a_n K(\vec{r};\vec{r}_n).$$

(3.37)

Thus, the same representation is used in system (3.36) and in relation (3.37), and this representation is correct only in the domain outside the set $\bar{A}$ of the wave field singularities.

In the matrix notation, system (3.36) becomes

$$K\bar{a} = \bar{b}, \quad \bar{a} = K^{-1} \cdot \bar{b}$$
$$\bar{a} = \{a_n\}_{n=1}^N, \quad \bar{b} = \{b_m\}_{m=1}^N, \quad K = \{K_{nm}\}_{n,m=1}^N, \quad b_m = -U^0(\vec{r}_m), \tag{3.38}$$

where

$$K_{nm} = \left\{ H_0^{(2)}\left(k|\vec{r}_{\Sigma n} - \vec{r}'|\right) - \frac{W}{k\kappa(\varphi_n)} \left[\rho(\varphi_n) \frac{\partial H_0^{(2)}\left(k|\vec{r}_{\Sigma n} - \vec{r}'|\right)}{\partial r'} \right. \right.$$
$$\left. \left. - \frac{\rho'(\varphi_n)}{\rho(\varphi_n)} \frac{\partial H_0^{(2)}\left(k|\vec{r}_{\Sigma n} - \vec{r}'|\right)}{\partial \varphi'} \right] \right\} \Bigg|_{\vec{r}' = \vec{r}_n}.$$

We use the summation theorem

$$H_0^{(2)}\left(k|\vec{r} - \vec{r}_n|\right) = \sum_{p=-\infty}^{\infty} J_p(kr_n) H_p^{(2)}(kr) e^{ip(\varphi - \varphi_n)}, \quad r > r_1, \tag{3.39}$$

and relations (3.37) and (3.35) to obtain

$$u^1(\vec{r}) = \sum_{p=-\infty}^{\infty} \left(\sum_{n=1}^N a_n \left\{ J_p(kr_n) - \frac{W}{k\kappa(\varphi_n)} \left[kr_n J_p'(kr_n) + ip \frac{\rho'(\varphi_n)}{r_n} J_p(kr_n) \right] \right\} e^{-ip\varphi_n} \right)$$
$$H_p^{(2)}(kr) e^{ip\varphi} = \sum_{p=-\infty}^{\infty} c_p H_p^{(2)}(kr) e^{ip\varphi},$$

where

$$c_p = \sum_{n=1}^N a_n J_{pn}$$

and $J_{pn} = \left\{ J_p(kr_n) - \dfrac{W}{k\kappa(\varphi_n)} \left[kr_n J_p'(kr_n) + ip \dfrac{\rho'(\varphi_n)}{r_n} J_p(kr_n) \right] \right\} e^{-ip\varphi_n}.$

In the matrix notation, we have

$$\bar{c} = J \cdot \bar{a} = \left(J \cdot K^{-1} \right) \cdot \bar{b} = T \cdot \bar{b}, \tag{3.40}$$

where $T = J \cdot K^{-1}$ is a T-matrix.

For the scattering pattern, by analogy with (1.43) we have

$$g(\varphi) = \int_0^{2\pi} J(\varphi')\exp[ik\rho(\varphi')\cos(\varphi-\varphi')]\,d\varphi' = \sum_{n=-\infty}^{\infty} c_n i^n e^{in\varphi}. \qquad (3.41)$$

As in the traditional *T*-matrix method, the obtained relations allow us easily to averaging the characteristics of particle scattering over the orientation angles. For example, for the scattering pattern averaged over the orientation angles, (3.40) implies

$$\langle g(\varphi)\rangle = \sum_{n=-\infty}^{\infty} \langle c_n \rangle i^n e^{in\varphi}, \qquad (3.42)$$

and (see (3.40))

$$\langle \bar{c}\rangle = T \cdot \langle \bar{b}\rangle. \qquad (3.43)$$

If, for example,

$$U^0 = \exp(-ikr\cos(\varphi-\varphi_0)) = \sum_{n=-\infty}^{\infty} (-i)^n J_n(kr)e^{in(\varphi-\varphi_0)} \qquad (3.44)$$

is a plane wave and the particle orientation with respect to the radiation angles φ_0 is equiprobable, i.e., $w(\varphi_0) = 1/2\pi$, then, as follows from (3.38),

$$\langle b_n \rangle = -\frac{1}{2\pi}\int_0^{2\pi} \exp(-ikr_\Sigma(\alpha_n)\cos(\alpha_n-\varphi_0))\,d\varphi_0 = -J_0(kr_\Sigma(\alpha_n)). \qquad (3.45)$$

In contrast to the traditional *T*-matrix method, this approach can be applied to scatterers of any geometry with an analytic boundary. We illustrate this with some special examples.

We consider the problem of diffraction of the plane wave (3.44) incident at the angle $\varphi_0 = 0$ at an infinite elliptic cylinder with the Dirichlet boundary condition ($W = 0$). First, we consider the Rayleigh elliptic cylinder with semiaxes $ka = 8$, $kc = 11$. Figure 3.28 shows the scattering pattern on the logarithmic scale $(10 \cdot \lg(\frac{2}{\pi}|g(\varphi)|^2))$ for this problem, and Fig. 3.29 shows the pattern averaged over the radiation angles uniformly distributed on the interval $\varphi_0 \in [0, 2\pi]$.

The patterns were obtained by modified *T*-matrix method (MTMM) for $N = 50$ and $P = 20$ (P is the summation limit in expansion (3.44)). Both patterns graphically coincide with the corresponding patterns in Demin *et al.*

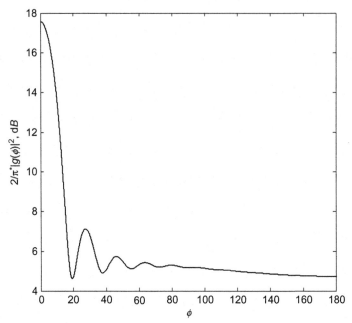

Figure 3.28 Scattering pattern for elliptic cylinder with semiaxes $ka = 8$, $kc = 11$.

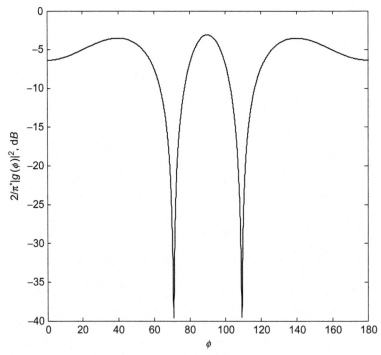

Figure 3.29 Pattern of elliptic cylinder with semiaxes $ka = 8$, $kc = 11$ averaged over radiation angles.

[83], where they were calculated by the classical *T*-matrix method and the pattern equation method (PEM) [94]. However, it should be noted that the computational algorithm based on the discrete sources is much more convenient and stable in practice than the TMM and PEM algorithms, because it does not need calculations of integrals and special functions of higher orders.

Now let us study the accuracy of the method in the considered example more precisely. We estimate the accuracy according to the following criteria, namely, by the discrepancy value and the rate of its decrease with increasing size of the algebraic system and by the accuracy of the optical theorem and the accuracy of construction of the scattering pattern.

We compare the MTMM scattering pattern $g_N(\varphi)$ with the pattern $g^s(\varphi)$ considered as "standard". As the "standard" pattern we take the pattern calculated with a high accuracy (the minimal possible discrepancy is 10^{-15}) by the NFM with the singularities of the field analytic continuation taken into account. The accuracy of calculating the pattern compared with the standard pattern is calculated in the linear metric using the formula

$$\Delta g^{\mathrm{max}} = \max \left| \frac{|g_N(\varphi)|}{\max|g_N(\varphi)|} - \frac{|g^s(\varphi)|}{\max|g^s(\varphi)|} \right|. \tag{3.46}$$

To estimate the error of the numerical solution, we calculate the discrepancy of the boundary condition at points between the collocation points. The discrepancy value directly depends on the number of collocation points N and if this number is insufficient, then the discrepancy value is large. The rate of the discrepancy decrease with increasing approximation level N allows us to estimate the rate of the internal convergence of the computational algorithm. The discrepancy value can also be calculated in the linear metric by the formula

$$\Delta_N(\varphi_m) = |K(\varphi_m) \cdot \bar{a} - \bar{b}(\varphi_m)|, \tag{3.47}$$

where $\varphi_m = \frac{2\pi}{N}(m-1)$.

Since the accuracy of the pattern calculation can be estimated only with respect to the "standard" pattern, as an additional criterion of the pattern calculation accuracy, we verify the optical theorem in the case where the initial field is a plane wave. In the two-dimensional case, the optical theorem can be written as

$$\int_0^{2\pi} |g(\varphi)|^2 d\varphi = 2\pi Reg(\varphi = \varphi_0). \tag{3.48}$$

To estimate the accuracy of the optical theorem, we calculate the quantity

$$\Delta_{\text{ot}} = \left| 2\pi Reg(\varphi = \varphi_0) - \int_0^{2\pi} |g(\varphi)|^2 d\varphi \right|. \qquad (3.49)$$

Table 3.2 illustrates the results of calculations of the quantities $\Delta g^{\max}$, $\max\Delta_N = \Delta_N^{\max}$, Δ_{ot}, and the time of calculation for different N by two methods of construction of the curve Σ: (1) S is analytically deformed to the singularities (MTMM), and (2) Σ is a circle of radius $kr_m = 7.55$ as in the standard T-matrix method with discrete sources (TMMDS) [29]. Table 3.3 shows the values of $\Delta g^{\max}, \Delta_{\text{ot}}$, and the time of calculation for different M by the classical TMM ($2M + 1 \times 2M + 1$ is the size of the system of algebraic equations in TMM).

It follows from Tables 3.2 and 3.3 that using the analytic deformation technique, we can obtain a higher rate of the computational algorithm convergence. Although the use of a circle as Σ is correct in this case, the results of the classical TMM and TMMDS are significantly worse than the results obtained by MTMM, and TMMDS has the greatest error. It follows from Table 3.3 that TMM has the following two algorithmic drawbacks, namely, the computation time is much greater than that in MTMM and the solution is destroyed completely starting from some M.

Now we consider the elliptic cylinder, which does not satisfy the Rayleigh hypothesis, with semiaxes $ka = 2$, $kc = 10$ for $\varphi_0 = \pi$. Figure 3.30 presents the scattering pattern on the logarithmic scale for the considered problem, and Fig. 3.31 shows the pattern averaged over the radiation angles uniformly distributed on the interval $\varphi_0 \in [0, 2\pi]$. The patterns are obtained by MTMM for $N = 200$ and $P = 20$. Both graphs coincide with the patterns in Demin et al. [83] and the patterns obtained by PEM, while the classical TMM, as shown in Demin et al. [83], gives a false result in this case.

For the considered example, Table 3.4 shows the results of calculations of $\Delta g^{\max}$, $\min\Delta_N = \Delta_N^{\min}$, $\Delta_N^{\max}$, and Δ_{ot} for different N by two methods for constructing the curve Σ: (1) S is analytically deformed to the singularities (MTMM), (2) Σ is obtained by decreasing the radius vector circumscribing S by the value $k\delta = 0.1$. The discrepancies have overshoots near the singularities (see Fig. 3.32), and, therefore, we here present not only the maximal but also the minimal values of the discrepancies. Table 3.5 shows the values of $\Delta g^{\max}$ and Δ_{ot} for different M in the classical TMM. The results obtained by

Table 3.2 Quantities Δg^{max}, $max\Delta_N = \Delta_N^{max}$, Δ_{ot}, and the time of calculation for different N

N	Analytic deformation of S till the singularities			Circle surrounding the singularities ($kr_m = 7.55$)			t, s
	Δg^{max}	Δ_N^{max}	Δ_{ot}	Δg^{max}	Δ_N^{max}	Δ_{ot}	
15	0.76707	0.15488	9.1264	0.66196	1.3663	16.4059	0.568140
20	0.14164	0.016738	0.1351	0.19432	0.24029	1.4921	0.616969
25	0.013194	0.00029976	0.0073298	0.083578	0.12893	0.42563	0.622717
30	0.0013849	1.9173×10^{-5}	0.000363	0.032776	0.067347	0.048168	0.641958
35	0.00010996	5.9275×10^{-7}	9.7121×10^{-6}	0.014983	0.038114	0.0026443	0.719283
50	2.4096×10^{-8}	5.2738×10^{-11}	9.8577×10^{-10}	0.0036916	0.010217	0.0066815	0.725088
100	6.3984×10^{-12}	3.7943×10^{-15}	7.1054×10^{-15}	9.2369×10^{-5}	0.0002589	0.0002603	1.375544

Table 3.3 Quantities Δg^{max}, $\max\Delta_N = \Delta_N^{max}$, Δ_{ot}, and the time of calculation for different N in the classical TMM

M	Δg^{max}	Δ_{ot}	t, s
15	0.001093	0.0047652	4.570527
20	1.7207×10^{-5}	5.1425×10^{-5}	7.573743
25	1.5947×10^{-7}	1.4195×10^{-7}	8.964420
30	1.2098×10^{-9}	1.5153×10^{-9}	12.803415
35	3.4189×10^{-7}	1.9271×10^{-5}	12.953874
50	6.3798×10^{-7}	4.1215×10^{-6}	16.293146
100	0.75283	9.1505×10^{3}	42.557182

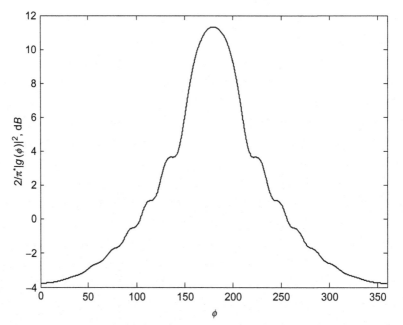

Figure 3.30 Scattering pattern for elliptic cylinder with semiaxes $ka = 2$, $kc = 10$.

TMMDS are approximately the same as in TMM, and hence we do not present them here.

The results shown in Table 3.4 are calculated in both cases with the surrounded singularities of the analytic continuation, but MTMM permits attaining an accuracy of the order of 10^{-7}, while MTMM without any analytic deformation permits obtaining only 10^{-3} for the same N. But as one can

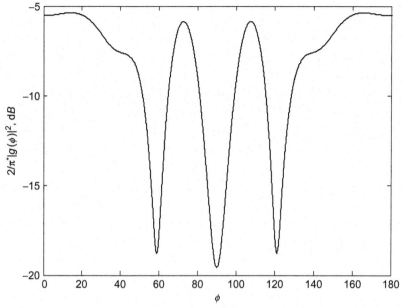

Figure 3.31 Pattern for elliptic cylinder with semiaxes $ka = 2$, $kc = 10$ averaged over radiation angles.

see in Table 3.5, the classical TMM does not work at all, i.e., the error of the pattern $\Delta g^{\max}$ is close to its maximal value 1 and does not decrease with increasing M.

Figure 3.32 illustrates the discrepancies $\Delta_N(\varphi_m)$ obtained by MTMM for different N: curves 1-6 correspond to the value of N in Table 3.4 (curve 1 corresponds to $N = 50$, ..., and curve 6 corresponds to $N = 500$). One can see that the computational algorithm of MTMM has a high accuracy even for such an oblate scatterer.

We consider the problem of plane wave diffraction at a cylinder whose cross-section has the shape of a 10-leaf multifoil ($r(\varphi) = b + a\cos 10\varphi$) with the parameters $ka = 5$, $kc = 10$ for $\varphi_0 = \pi/2$. Here as in the preceding example, we again solve the problem with the Dirichlet boundary condition. We compare the results obtained by MTMM and by the classical TMM. Figure 3.33 shows the scattering patterns on the logarithmic scale for this problem (curve 1 is obtained by MTMM for $N = 400$ and $P = 20$, and curve 2 is obtained by TMM for $M = 40$). Figure 3.34 shows the pattern averaged over the radiation angles uniformly distributed on the interval $\varphi_0 \in [0, 2\pi]$, which is obtained by MTMM. One can see that TMM again gives a false

Table 3.4 Quantities Δg^{max}, $\min\Delta_N = \Delta_N^{min}$, Δ_N^{max}, and Δ_{ot} obtained by two methods for constructing the curve Σ till the singularities ($k\delta=0.1$)

N	Analytic deformation of S till the singularities				Nonanalytic deformation of S till the singularities			
	Δg^{max}	Δ_N^{min}	Δ_N^{max}	Δ_{ot}	Δg^{max}	Δ_N^{min}	Δ_N^{max}	Δ_{ot}
50	0.026654	0.003345	0.43487	0.018277	0.23353	0.005065	0.91801	0.13131
100	0.000689	0.000224	0.023231	0.000428	0.028207	2.953×10^{-5}	0.29848	0.064146
200	4.953×10^{-6}	2.061×10^{-6}	0.000206	2.439×10^{-6}	0.007311	5.958×10^{-8}	0.073456	0.016597
300	7.673×10^{-7}	2.536×10^{-8}	2.508×10^{-6}	3.031×10^{-7}	0.002391	2.635×10^{-10}	0.015898	0.005841
400	7.783×10^{-7}	3.510×10^{-10}	3.434×10^{-8}	2.814×10^{-7}	0.001236	1.230×10^{-12}	0.58825	0.002016
500	7.785×10^{-7}	5.184×10^{-12}	5.018×10^{-10}	2.811×10^{-7}	0.001172	4.857×10^{-5}	0.75648	0.000459

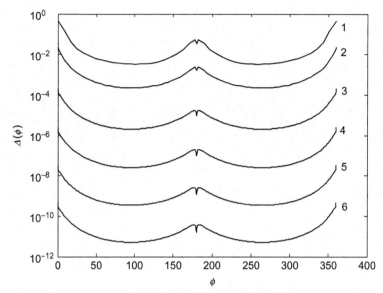

Figure 3.32 Discrepancies $\Delta_N(\varphi_m)$ obtained by MTMM for different N.

Table 3.5 Values of Δg^{max} and Δ_{ot} for different M in the classical TMM

M	Δg^{max}	Δ_{ot}
15	0.77283	55.182634
20	0.52759	38.411849
25	0.48969	29.055242
30	0.77326	1.8036×10^3
50	0.82231	4.2002×10^2
100	0.82162	1.4156×10^7
150	0.82235	1.0941×10^7

result, because the 10-leaf multifoil under study does not satisfy the Rayleigh hypothesis.

One can also see that the meaned scattering pattern reflects in a sense the scatterer geometry (it has 10 equal leaves).

For the considered example, Table 3.6 shows the results of calculations of Δg^{max}, Δ_N^{min}, Δ_N^{max}, and Δ_{ot} for different N by the following two methods for constructing the curve Σ: (1) S is analytically deformed till the singularities

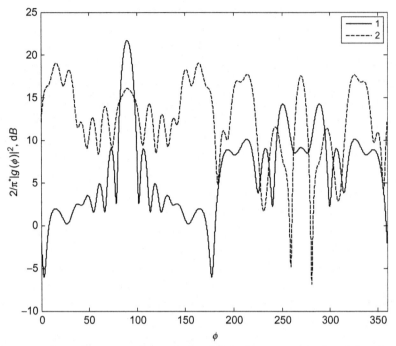

Figure 3.33 Scattering pattern for 10-leaf multifoil with parameters $ka = 5$, $kc = 10$.

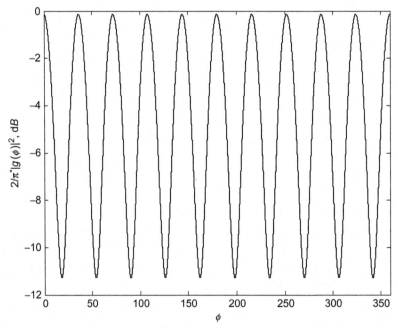

Figure 3.34 Pattern for 10-leaf multifoil with parameters $ka = 5$, $kc = 10$ averaged over radiation angles.

Table 3.6 Quantities Δg^{max}, Δ_N^{min}, Δ_N^{max}, and Δ_{ot} for 10-leaf multifoil

N	Analytic deformation of S till the singularities				Circle of radius kr_m=4.95			
	Δg^{max}	Δ_N^{min}	Δ_N^{max}	Δ_{ot}	Δg^{max}	Δ_N^{min}	Δ_N^{max}	Δ_{ot}
50	0.21238	0.13738	3.5223	1.3722	0.78525	3.013×10^{-6}	1.156×10^{-5}	25.7
100	0.15496	0.000135	0.27227	0.12809	0.76973	2.002×10^{-16}	2.301×10^{-14}	53.4
200	0.007518	1.915×10^{-8}	0.00643	0.000917	0.81958	1.062×10^{-14}	3.496×10^{-13}	87559.4
300	0.000327	3.175×10^{-12}	0.00022	6.3856×10^{-5}	0.95605	1.144×10^{-15}	4.845×10^{-14}	1395.2
400	1.482×10^{-5}	2.524×10^{-14}	7.77×10^{-6}	3.5787×10^{-6}	0.87568	1.233×10^{-15}	3.837×10^{-14}	1123.3
500	6.847×10^{-7}	2.989×10^{-16}	3.03×10^{-7}	1.9214×10^{-7}	0.93662	5.901×10^{-16}	5.59×10^{-14}	2324

(MTMM), (2) Σ is a circle of radius $kr_m = 4.95$ (TMMDS). Table 3.7 shows the values of $\Delta g^{\max}$ and Δ_{ot} for different M obtained by TMM.

The results in these tables are similar to the preceding ones. Although the considered 10-leaf multifoil is not an extended figure, it does not satisfy the Rayleigh hypothesis, and therefore neither TMM nor TMMDS give true results.

In conclusion, we consider the problem of plane wave diffraction at a cylinder with cross-section shaped as the superellipse ($\frac{x^{2m}}{c^{2m}} + \frac{y^{2m}}{d^{2m}} = 1$) with the parameters $ka = 2, kc = 10, m = 8$ for $\varphi_0 = \pi$. Figure 3.35 shows the

Table 3.7 Values $\Delta g^{\max}$ and Δ_{ot} for different M obtained by TMM

M	$\Delta g^{\max}$	Δ_{ot}
20	0.82717	3.1398×10^2
30	0.92526	2.4620×10^3
50	0.91122	2.0905×10^3
100	0.78854	8.9886×10^2
200	0.8785	4.1433×10^4

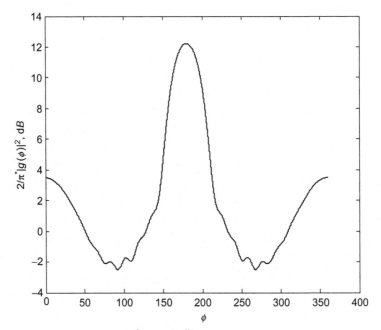

Figure 3.35 Scattering pattern for superellipse.

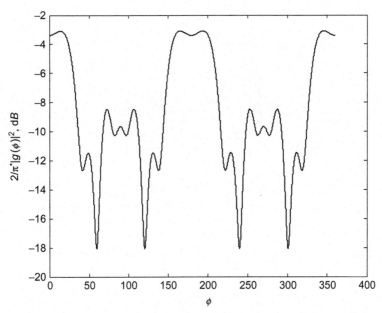

Figure 3.36 Scattering pattern for superellipse averaged over radiation angles.

scattering pattern on the logarithmic scale for the considered problem which is obtained by MTMM for $N = 400$ and $P = 20$. Figure 3.36 shows the pattern meaned over the radiation angles uniformly distributed on the interval $\varphi_0 \in [0, 2\pi]$, which is also obtained by MTMM. In this case, TMM does not permit obtaining true results, as follows from Table 3.8.

For the considered example, Table 3.8 shows the results of calculations of $\Delta g^{\max}$, $\Delta_N^{\min}$, $\Delta_N^{\max}$, and Δ_{ot} for different N by the following two methods for constructing the curve Σ: (1) S is analytically deformed till the singularities (MTMM), and (2) Σ is obtained be decreasing the radius vector of S by $k\delta = 0.01$. Table 3.9 shows the values of $\Delta g^{\max}$ and Δ_{ot} obtained by the classical TMM for different M. In TMMDS, we obtain approximately the same results as in TMM, and, therefore, we do not present them here. The results shown in Tables 3.8 and 3.9 are similar to the results shown in Tables 3.4 and 3.5 for the non-Rayleigh ellipse. Despite the fact that the singularities of the analytic continuation for this superellipse are very close to its boundary, MTMM again permits obtaining an accuracy of the order of 10^{-7}.

Table 3.8 Quantities Δg^{max}, $\min\Delta_N = \Delta_N^{min}$, Δ_N^{max}, and Δ_{ot} for non-Rayleigh ellipse

N	Analytic deformation of S to the singularities				Nonanalytic deformation of S to the singularities ($k\delta = 0.01$)			
	Δg^{max}	Δ_N^{min}	Δ_N^{max}	Δ_{ot}	Δg^{max}	Δ_N^{min}	Δ_N^{max}	Δ_{ot}
100	0.1188	0.0013357	0.21549	0.11351	0.21103	0.001025	0.77349	0.013733
200	0.002469	0.000136	0.022388	0.002474	0.049434	0.001183	0.39826	0.005973
300	0.000346	2.0534×10^{-5}	0.003728	0.000258	0.02588	0.000625	0.25497	0.002902
400	5.476×10^{-5}	3.468×10^{-6}	0.000650	3.373×10^{-5}	0.019475	0.000304	0.19061	0.000836
500	9.54×10^{-6}	6.121×10^{-7}	0.000117	5.4236×10^{-6}	0.020091	0.000150	0.14775	0.001213
700	3.279×10^{-7}	2.176×10^{-8}	4.102×10^{-6}	1.697×10^{-7}	0.057617	3.407×10^{-5}	0.18573	0.019945

Table 3.9 Values of Δg^{max} and Δ_{ot} for non-Rayleigh ellipse

M	Δg^{max}	Δ_{ot}
20	0.63408	24.7311
30	0.81316	89.5491
50	0.80027	1.4024×10^3
100	0.47947	1.5404×10^6
150	0.63444	1.4162×10^6

Thus, the proposed modification of the T-matrix method significantly extends the class of geometries for which a T-matrix can be obtained and a convenient stable high-convergence computational algorithm based on discrete sources can be used.

CHAPTER 4

Method of Continued Boundary Conditions

One of the most popular methods for solving the boundary-value problems for elliptic equations is their reduction to integral equations of the first or second type [5,8,11,12] and their subsequent algebraization. The kernels of such equations usually are fundamental solutions of the corresponding partial differential equations and (or) their derivatives and thus have singularities if the arguments coincide. This fact leads to a series of difficulties in the numerical realization of some algorithms for solving these equations.

The approach presented below allows us to overcome such difficulties and construct simple and efficient algorithms for solving the corresponding boundary-value problems [13,14,21,22,73,84–88].

4.1 METHOD OF CONTINUED BOUNDARY CONDITIONS FOR SCALAR DIFFRACTION PROBLEMS

4.1.1 Statement of the Problem and the Method Idea

We illustrate the main idea of the considered method by an example of boundary-value problems for the Helmholtz equation. The main idea can be generalized up to technical details to other equations of elliptic type.

For definiteness, we consider the solution of the external boundary-value problem for the Helmholtz equation, i.e., the problem of diffraction at a compact scatterer occupying the spatial domain $\bar{D}$. In the mathematical statement of this problem, in the domain $D_e = \mathbb{R}^3 \backslash \bar{D}$ (or $\mathbb{R}^2 \backslash \bar{D}$), it is required to determine the solution U^1 of the Helmholtz equation

$$\Delta U^1 + k^2 U^1 = 0, \tag{4.1}$$

where $k = \omega \sqrt{\varepsilon \mu}$ is the wave number, $k \in \mathbb{C}$, $k \neq 0$; on the boundary S of the domain $\bar{D}$ this solution must satisfy some boundary conditions, for example, of the form

$$\left(\alpha U(\vec{r}) + \beta \frac{\partial U(\vec{r})}{\partial n} \right) \Bigg|_S = 0, \tag{4.2}$$

where $\alpha, \beta = \text{const}$, $U = U^0 + U^1$ is the complete field, U^0 is the incident (primary) field, $\frac{\partial}{\partial n}$ is the differentiation in the direction of the outer normal on S, and α, β are some prescribed quantities, and the solution must also satisfy condition (3.3) at infinity.

Thus, the desired function $U^1(\vec{r})$ is real analytic [5,6] in the domain D_e, and, therefore, it also approximately satisfies condition (4.2) in a neighborhood of the boundary S, lying in D_e. If the boundary is such that the function $U^1(\vec{r})$ can analytically be continued into the domain $\bar{D}$ [9,10], then this neighborhood occupies both sides of the boundary S.

The idea of the continued boundary condition method (CBCM) is to transfer the boundary condition from the scatterer surface S to an auxiliary surface S_δ drawn inside the domain D_e at a certain sufficiently small distance δ from the surface S (see Fig. 4.1). In this case, the support of the auxiliary current generating the scattered field remains on the scatterer surface. As a result, instead of the exact statement of the boundary-value problem, we obtain an approximate problem whose solution has a prescribed error of the order of $k\delta$, which is demonstrated in Section 4.1.5.

Because any numerical solution of the boundary-value problem (4.1), (4.2) (including the solution based on a rigorous algorithm) is approximate, the use of approximate boundary conditions instead of exact conditions is quite admissible. Instead of rigorous statement (4.1), (4.2) of the boundary-value problem, we consider the problem with the boundary condition

$$\left(\alpha U(\vec{r}) + \beta \frac{\partial U(\vec{r})}{\partial n} \right) \Bigg|_{S_\delta} = 0, \tag{4.3}$$

where S_δ is the surface drawn inside the domain D_e at a sufficiently small distance δ from the surface S (see Fig. 4.1). Such a change of the statement

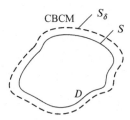

Figure 4.1 Geometry of problem.

of the problem allows us to avoid several serious computational difficulties. In particular, when the boundary-value problem (4.1), (4.3) is reduced to integral equations, the kernels of such equations have no singularities, and hence many computational problems related to the singularities contained in the equation kernels are removed.

The main advantages of CBCM are its universality and simplicity, and the CBCM universality is manifested first in the absence of any constraints on the scatterer geometry (it can be used both for scatterers with broken boundaries and for thin screens) and second in the possibility of reducing the boundary-value problem to integral equations of the first and second type with smooth kernel. Moreover, the CBCM proposes a uniform approach to solving the boundary-value problems independently of their type, the scatterer surface geometry, and the scattered field nature.

4.1.2 Derivation of CBCM Integral Equations

The CBCM integral equations are obtained according to the same scheme as the singular integral equations. For definiteness, we assume that the boundary S is piecewise smooth. As S_δ we take a piecewise smooth surface containing S and lying at a sufficiently small distance δ from it. Then the solution of problem (4.1), (4.3) can be reduced to a Fredholm integral equation of the first or second kind. The choice of the equation is dictated by the convenience reasons and depends on whether the equation has a simpler form for a given boundary condition. We discuss this below in more detail.

Now we use the following representation for the solution of the Helmholtz equation in the domain D_e (see Chapter 1):

$$U^1(\vec{r}) = \int_S \left\{ U(\vec{r}') \cdot \frac{\partial G_0}{\partial n'} - \frac{\partial U(\vec{r}')}{\partial n'} \cdot G_0(\vec{r},\vec{r}') \right\} ds', \qquad (4.4)$$

where $G_0(\vec{r},\vec{r}') = \frac{1}{4\pi} \cdot \frac{\exp(-i \cdot k \cdot |\vec{r}-\vec{r}'|)}{|\vec{r}-\vec{r}'|}$ is the fundamental solution of the Helmholtz equation in $\mathbb{R}^3$ ($G_0(\vec{r},\vec{r}') = \frac{1}{4i}H_0^{(2)}(k \cdot |\vec{r}-\vec{r}'|)$ in $\mathbb{R}^2$).

To obtain an integral equation of the first kind, we substitute expression (4.4) into the boundary condition (4.3) thus drawing the observation point M, which is determined by the radius vector $\vec{r}$, to the surface S_δ:

$$\alpha \int_S \left\{ U(\vec{r}') \cdot \frac{\partial G_0}{\partial n'} - \frac{\partial U(\vec{r}')}{\partial n'} \cdot G_0(\vec{r},\vec{r}') \right\} ds'$$

$$+ \beta \int_S \left\{ U(\vec{r}') \cdot \frac{\partial^2 G_0}{\partial n'\partial n} - \frac{\partial U(\vec{r}')}{\partial n'} \cdot \frac{\partial G_0}{\partial n} \right\} ds' = -\alpha U^0(\vec{r}) - \beta \frac{\partial U^0(\vec{r})}{\partial n}'$$

(4.5)

or

$$\int_S \left\{ U(\vec{r}') \cdot \left(\alpha \frac{\partial G_0}{\partial n'} + \beta \frac{\partial^2 G_0}{\partial n'\partial n} \right) - \frac{\partial U(\vec{r}')}{\partial n'} \cdot \left(\alpha G_0(\vec{r},\vec{r}') + \beta \frac{\partial G_0}{\partial n} \right) \right\} ds'$$

$$= -\alpha U^0(\vec{r}) - \beta \frac{\partial U^0(\vec{r})}{\partial n}.$$

Here $\vec{r}$, $\vec{r}'$ are the radius vectors of observation points $M(\vec{r}) \in S_\delta$ and integration points $N(\vec{r}') \in S$, respectively.

Then we take the boundary condition (4.3) into account in the integrand in Equation (4.5). Depending on the type of the boundary condition, either one of the functions $U(\vec{r})$ and $\frac{\partial U(\vec{r}')}{\partial n'}$ is zero on S_δ or both of these functions can linearly be expressed in terms of each other on S_δ.

Thus, in the case of a boundary condition of the first type, we obtain the Fredholm integral equation of the first kind:

$$\int_S \frac{\partial U(\vec{r}')}{\partial n'} \cdot G_0(\vec{r},\vec{r}') ds' = U^0(\vec{r})$$

(4.6)

for the unknown function $\frac{\partial U(\vec{r}')}{\partial n'} = I(\vec{r}')|_S$, which is an auxiliary current on the scatterer surface. Here and hereafter, we assume that $M(\vec{r}) \in S_\delta$ and $N(\vec{r}') \in S$.

For a boundary condition of the second type, we have the integral equation

$$\int_S U(\vec{r}') \cdot \frac{\partial^2 G_0(\vec{r},\vec{r}')}{\partial n'\partial n} ds' = -\frac{\partial U^0(\vec{r})}{\partial n}$$

(4.7)

for the unknown function $U(\vec{r})$, which is an auxiliary current on the scatterer surface.

Finally, for a boundary condition of the third type, we obtain

$$\int_S \frac{\partial U(\vec{r}')}{\partial n'} \cdot \left(\alpha G_0 + \beta \left(\frac{\partial G_0}{\partial n'} + \frac{\partial G_0}{\partial n} \right) + \frac{\beta^2}{\alpha} \frac{\partial^2 G_0}{\partial n'\partial n} \right) ds'$$

$$= \alpha U^0(\vec{r}) + \beta \frac{\partial U^0(\vec{r})}{\partial n}.$$

(4.8)

To obtain the integral equation of the second kind, we substitute representation (4.4) into the identity $U = U^1 + U^0$ and obtain

$$U(\vec{r}) - \int_S \left\{ U(\vec{r}') \cdot \frac{\partial G_0}{\partial n'} - \frac{\partial U(\vec{r}')}{\partial n'} \cdot G_0(\vec{r}, \vec{r}') \right\} ds' = U^0(\vec{r}). \quad (4.9)$$

Then, as in the preceding case, we take the boundary conditions (4.3) into account in (4.9). For the boundary condition of the first type, after differentiation along the outer normal on S, we obtain

$$\frac{\partial U(\vec{r})}{\partial n} + \int_S \frac{\partial U(\vec{r}')}{\partial n'} \cdot \frac{\partial G_0(\vec{r}, \vec{r}')}{\partial n} ds' = \frac{\partial U^0(\vec{r})}{\partial n}, \quad (4.10)$$

for the boundary condition of the second type,

$$U(\vec{r}) - \int_S U(\vec{r}') \cdot \frac{\partial G_0}{\partial n'} ds' = U^0(\vec{r}), \quad (4.11)$$

and for the boundary condition of the third type,

$$-\frac{\beta}{\alpha} \frac{\partial U(\vec{r})}{\partial n} + \int_S \frac{\partial U(\vec{r}')}{\partial n'} \left(\frac{\beta}{\alpha} \frac{\partial G_0}{\partial n'} + G_0(\vec{r}, \vec{r}') \right) ds' = U^0(\vec{r}). \quad (4.12)$$

It may seem that Equations (4.10)–(4.12) are not equations of the second kind, because S and S_δ are different. But this is not true. Indeed, let the surfaces S and S_δ be given, for example, in spherical coordinates by the equations $r'|_S = r(\theta', \varphi')$, $r|_{S_\delta} = \rho(\theta, \varphi)$, $\theta', \theta \in [0, \pi], \varphi', \varphi \in [0, 2\pi]$. Then, under the assumption that $\frac{\partial U(\vec{r})}{\partial n}|_{S_\delta} \cong \frac{\partial U(\vec{r}')}{\partial n'}|_S = I(\theta, \varphi)$ for $\theta' = \theta, \varphi' = \varphi$, we obtain a Fredholm integral equation of the second kind for the function $I(\theta, \varphi)$.

Equations (4.10)–(4.12) differ from the equations obtained by the traditional method of current integral equations (MCIE) in that there is no multiplier $1/2$ before the separately standing unknown function, which appears because of the double-layer potential jump on the scatterer surface. It is obvious that such a jump cannot be at the distance δ from the scatterer on which the boundary condition is posed. But in numerical calculations, this multiplier must be added (see below).

We see that the kernel of an integral equation of the first kind has a simpler form for boundary condition of the first type and the kernels of integral equations of the second kind have a simpler form for boundary conditions of the second and third type. We also note that all obtained equations have smooth kernels.

4.1.3 Existence and Uniqueness of the CBCM Integral Equation Solution

For Equations (4.6)–(4.8), the existence and uniqueness theorems can be proved as in Chapter 2. For example, for Equation (4.6), these theorems are formulated as follows:

1. *Assume that a simple closed surface S satisfies the condition that k is not an eigenvalue of the internal homogeneous Dirichlet problem for a domain inside S. Then Equation (4.6) is solvable if, and only if, S surrounds all singularities of the solution* $U_\delta^1\left(\vec{r}\right)$ *of the boundary-value problem* (4.1), (4.3), *where* $\alpha = 1$, $\beta = 0$.
2. *If the conditions of Theorem 1 are satisfied, Equation (4.6) has a unique solution.*

The proof of these theorems repeats almost word for word the proof given in Chapter 2.

4.1.4 Well-Posedness of the Numerical Solution of the CBCM Integral Equation

As shown in Section 4.1.2, CBCM permits reducing the boundary-value problem to a Fredholm integral equation of the second kind with a smooth kernel independently of the type of the boundary condition. To solve such a problem numerically is always a well-posed problem, i.e., its solution always (except for the cases where the parameter of the integral equation coincides with an eigenvalue of the kernel) exists and is unique and stable (a small variation in the solution of this equation corresponds to small variations in the right-hand side of the equation).

But in several cases, for example, in the case of a boundary condition of the first type, it is more convenient to reduce the boundary-value problem to a Fredholm equation of the first kind. This equation has a smooth kernel, and as is known, to solve this numerically is an ill-posed problem [89]. The above-listed theorems state that, under certain conditions, such a problem has a unique solution, but this solution can be unstable.

Nevertheless, this fact is not of principal importance in our case. The point is that all quantities that we are interested in can be treated as functionals of the sought currents on S. In particular, if we are interested in the current distribution on S, then, instead of this current, we can calculate (with a prescribed accuracy) the corresponding quantity on the surface S_δ^2 located at an arbitrarily small distance from S_δ, and this calculation is already quite stable [73]. We illustrate this fact by a special example.

Let us consider the problem of a plane wave (see formula (3.21)) diffraction at an oblate spheroid with semiaxes $ka = 5$, $kc = 0.125$, and $k\delta = 10^{-4}$. The scattering pattern for the considered body is close to the pattern for a

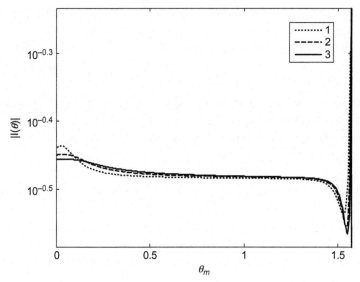

Figure 4.2 Current on the surface S_δ^2.

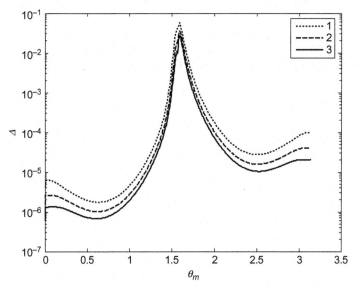

Figure 4.3 Boundary condition discrepancy on the surface S_δ^2.

disk of the same radius, which in given in Ref. [132]. Figures 4.2 and 4.3 show the respective values of the current and the boundary condition discrepancy on the surface S_δ^2 for different N: 1—$N = 300$, 2—$N = 500$, and 3—$N = 700$. One can see that, despite a sufficiently large size of the algebraic system, the calculations of the value of the current are quite stable. Figure 4.3 shows that the boundary condition is satisfied in this case with a rather high accuracy of the order of 10^{-4} (except for the discrepancy overshoot at $\theta = \pi$, which can be explained by the geometry of the considered body).

Moreover, the instability of calculations of the current distribution on S is manifested only if the size of the corresponding system of algebraic equations becomes extremely large. If the choice of the relation between the size of the system of algebraic equations N and the value of the parameter δ is correct (see Section 4.3.2), then all calculations are quite stable. This is confirmed by the results of similar studies given in Ref. [90]. As an example, we refer to Fig. 6 in that paper, which illustrates the results of calculations of the current on the cylindrical screen for $ka = 120$ and the discrepancy. This figure shows that the exact and approximate solutions (currents) graphically coincide and the approximate solution has no oscillations. This is due to the correct choice of the system size. In this case, the discrepancy of the boundary condition is good, and hence, it does not make sense significantly to increase the size of the system.

4.1.5 CBCM Rigorous Solution of Some Diffraction Problems and Estimation of the Error of the Method

To estimate the error obtained by passing to continued boundary conditions, we consider the boundary surface for which the approximately posed boundary-value problem (4.1), (4.3) can be solved exactly. In the three-dimensional case, such surfaces are the sphere and spheroid (for boundary conditions of the first or second type).

Now we determine the solution of the problem of diffraction of a plane wave incident at angles φ_0, $\theta_0 = 0$ at a sphere of radius a, where the Dirichlet boundary condition is satisfied, in the exact and approximate statement. As S_δ we take a sphere of radius $a + \delta$. The CBCM integral equation for this problem has the form (4.6).

First, we consider the problem in the exact statement. For this, in (4.6), it is necessary to assume that the surfaces S and S_δ coincide ($\delta = 0$). We pass to spherical coordinates and expand the unknown function, the kernel, and the right-hand side of Equation (4.6) in series in spherical functions:

$$I(\theta', \varphi') = \sum_{n=0}^{\infty} \sum_{m=-n}^{n} a_{nm} P_n^m(\cos\theta') e^{im\varphi'}, \qquad (4.13)$$

$$\frac{i}{k}G_0\left(\vec{r},\vec{r}'\right)$$

$$=\sum_{\nu=0}^{\infty}\sum_{\mu=-\nu}^{\nu}(2\nu+1)\frac{(\nu-\mu)!}{(\nu+\mu)!}h_\nu^{(2)}(kr)j_\nu(kr')P_\nu^\mu(\cos\theta)P_\nu^\mu(\cos\theta')e^{i\mu(\varphi-\varphi')}, \tag{4.14}$$

$$U^0(\vec{r})=\sum_{n=0}^{\infty}\sum_{m=-n}^{n}(-i)^n(2n+1)\frac{(n-m)!}{(n+m)!}j_n(kr)P_n^m(\cos\theta)P_n^m(\cos\theta_0)e^{im(\varphi-\varphi_0)}. \tag{4.15}$$

We substitute these expansions in Equation (4.6), take into account the expression

$$\int_0^\pi P_n^m(\cos\theta')P_\nu^m(\cos\theta')\sin\theta'\,d\theta'=\frac{2}{2n+1}\frac{(n+m)!}{(n-m)!}\delta_{n\nu},$$

and obtain the values of the coefficients a_{nm}:

$$a_{nm}=-(-i)^n\frac{P_n^{(m)}(\cos\theta_0)e^{-im\varphi_0}(2n+1)(n-m)!}{4\pi h_n^{(2)}(ka)(n+m)!}. \tag{4.16}$$

If $\theta_0=0$, then $P_n^{(m)}(1)=\delta_{m0}$ and

$$a_{nm}=a_n\delta_{m0}=-(-i)^n\frac{2n+1}{4\pi h_n^{(2)}(ka)}\delta_{m0}. \tag{4.17}$$

Then we have

$$I(\theta',\varphi')=-\sum_{n=0}^{\infty}\frac{(-i)^n(2n+1)}{4\pi h_n^{(2)}(ka)}P_n(\cos\theta') \tag{4.18}$$

Similarly, we obtain the solution of the problem in the approximate statement ($\delta\neq0$ in (4.6)). The unknown function can be represented as

$$I_\delta(\theta',\varphi')=\sum_{n=0}^{\infty}\sum_{m=-n}^{n}a_{nm}^\delta P_n^{(m)}(\cos\theta')e^{im\varphi'}. \tag{4.19}$$

For the other functions we use the same representations as above and, for $\theta_0=0$, we obtain

$$a_{nm}^\delta=a_n^\delta\delta_{m0}=-(-i)^n\frac{(2n+1)j_n(ka+k\delta)}{4\pi j_n(ka)h_n^{(2)}(ka)}\delta_{m0},$$

$$I_\delta(\theta',\varphi')=-\sum_{n=0}^{\infty}\frac{(-i)^n(2n+1)j_n(ka+k\delta)}{4\pi j_n(ka)h_n^{(2)}(ka+k\delta)}P_n(\cos\theta'). \tag{4.20}$$

Now we calculate the mean square error

$$\Delta(k\delta) = \frac{1}{4\pi} \int_0^{2\pi} \int_0^{\pi} |I(\theta', \phi') - I_\delta(\theta', \phi')|^2 \sin\theta' d\theta' d\phi' = \sum_{n=0}^{\infty} \frac{1}{2n+1} |a_n - a_n^\delta|^2$$

to determine the solution error, which arose due to the use of the approximate boundary conditions.

We transform this expression for the difference of the coefficients taking into account that

$$j_n'(ka)h_n^{(2)}(ka) - j_n(ka)h_n'^{(2)}(ka) = \frac{i}{(ka)^2},$$

$$a_n - a_n^\delta = \frac{(-i)^n(2n+1)ik\delta}{4\pi(ka)^2 j_n(ka)h_n^{(2)}(ka)h_n^{(2)}(ka+k\delta)},$$

$$a_n - a_n^\delta = \frac{(-i)^n(2n+1)ik\delta}{2\pi^2 ka \sqrt{\dfrac{\pi}{2(ka+k\delta)}} J_{n+(1/2)}(ka)H_{n+(1/2)}^{(2)}(ka)H_{n+(1/2)}^{(2)}(ka+k\delta)}.$$

The error can be estimated up to 10^{-5} as

$$\Delta(k\delta) \sim \sum_{n=0}^{2ka} \frac{1}{2n+1} |a_n - a_n^\delta|^2. \tag{4.21}$$

We determine the error of the method when the scattering pattern is constructed. Substituting the obtained value of the current in the formula for the pattern

$$F(\theta, \varphi) = \int_0^{2\pi} \int_0^{\pi} I(\theta', \phi') \exp(ika(\sin\theta\sin\theta'\cos(\varphi - \phi') + \cos\theta\cos\theta'))$$
$$\sin\theta' d\theta' d\phi',$$

we obtain

$$F(\theta, \varphi) = \sum_{n=0}^{\infty} a_n 4\pi(-i)^n j_n(ka) \frac{1}{2n+1} P_n(\cos\theta)$$

for the exact boundary condition and

$$F_\delta(\theta, \varphi) = \sum_{n=0}^{\infty} a_n^\delta 4\pi(-i)^n j_n(ka) \frac{1}{2n+1} P_n(\cos\theta)$$

for the approximate boundary condition.

We determine the mean square error for the scattering pattern:

$$\Delta(k\delta) = \frac{1}{4\pi} \int_0^{2\pi} \int_0^{\pi} |F(\theta,\varphi) - F_\delta(\theta,\varphi)|^2 \sin\theta \, d\theta \, d\phi$$

$$= \sum_{n=0}^{\infty} \frac{1}{2n+1} \left| 4\pi j_n(ka) \frac{1}{2n+1} (a_n - a_n^\delta) \right|^2,$$

$$\Delta(k\delta) = \sum_{n=0}^{\infty} \frac{1}{2n+1} |A_n - A_n^\delta|^2, \tag{4.22}$$

where $A_n = 4\pi j_n(ka)\frac{1}{2n+1}a_n$ and $A_n^\delta = 4\pi j_n(ka)\frac{1}{2n+1}a_n^\delta$.

This error can be estimated as

$$\Delta(k\delta) \sim \sum_{n=0}^{2ka} \frac{1}{2n+1} |A_n - A_n^\delta|^2. \tag{4.23}$$

Figure 4.4a and b show how the error of the method depends on $k\delta$ for different values of ka for the current density and the pattern, respectively.

These figures show that, for the problem under study, the error arising when we pass to the approximate boundary conditions monotonically decreases with decreasing $k\delta$. In this case, the error of determining the current density for spheres of dimensions less than or equal to three wavelengths does not exceed $k\delta$. And the error of determining the scattering pattern is at least by an order of magnitude less than the error of determining the current density (for bodies of dimensions greater than 1.3 wavelengths). These results can be used as reference points in the choice of $k\delta$ depending on the body size and the required accuracy of the problem solution not only for the sphere but also for other smooth surfaces.

Now we determine the solution of the problem of diffraction of a plane wave at a prolate spheroid with the Neumann boundary condition satisfied on its surface [87] (in this case, the boundary-value problem reduces to Equation (4.11)) in the exact and approximate statements; we also estimate the error in a different metric.

We present several relations which will be used later [40] and operate in the coordinate system of the prolate spheroid. In these coordinates, we have

$$G_0(\vec{r}, \vec{r}') = \frac{k}{2\pi i} \sum_{n=0}^{\infty} \sum_{m=-n}^{n} \frac{S_{mn}(c,\eta')}{N_{mn}(c)} S_{mn}(c,\eta) R_{|m|n}^{(1)}(c,\xi') R_{|m|n}^{(4)}(c,\xi) e^{im(\varphi-\varphi')},$$

$$\tag{4.24}$$

$$U^0 = 2\sum_{n=0}^{\infty} \sum_{m=-n}^{n} (-i)^n \frac{S_{mn}(c,\eta_0)}{N_{mn}(c)} S_{mn}(c,\eta) R_{|m|n}^{(1)}(c,\xi) e^{im(\varphi-\varphi_0)}. \tag{4.25}$$

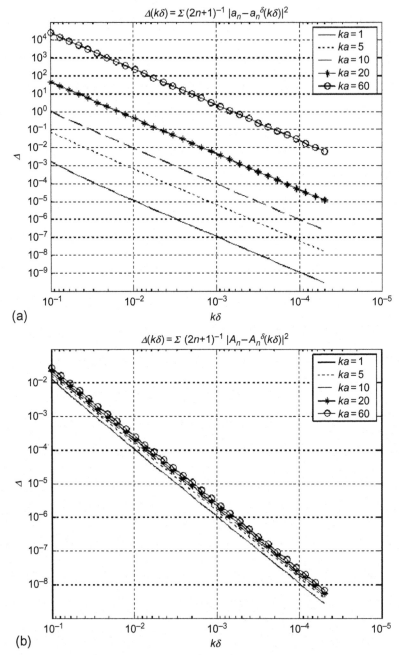

Figure 4.4 (a) Error of the method for the current density. (b) Error of the method for the pattern.

In the above formula, $S_{mn}(c, \eta)$ are angular prolate spheroidal functions of the first type of order m and of degree n, $R_{mn}^{(1),(4)}(c, \xi)$ are radial prolate spheroidal functions of the first (fourth) type, $N_{mn}(c) = 2 \sum\limits_{r=0,1}^{\infty}{}' \frac{(r+2m)!|d_r^{mn}(c)|^2}{(2r+2m+1)r!}$, where the prime on the sum symbol means that the sum is taken only over the even numbers r if $n - m$ is even, then the sum is taken only over odd numbers r and if $n - m$ is odd, then the coefficients $d_r^{mn}(c)$ are calculated by using the recursive relation

$$
\frac{(2m+r+2)(2m+r+1)c^2}{(2m+2r+3)(2m+2r+5)} d_{r+2}^{mn}(c) + \left[(m+r)(m+r+1) \right.
$$
$$
- \lambda_{mn}(c) + \frac{2(m+r)(m+r+1) - 2m^2 - 1}{(2m+2r-1)(2m+2r+3)} c^2 \left.\right] d_r^{mn}(c)
$$
$$
+ \frac{r(r-1)c^2}{(2m+2r-3)(2m+2r-1)} d_{r-2}^{mn}(c) = 0, \quad r \geq 0,
$$

where $\lambda_{mn}(c)$ are eigenvalues of the parameter λ of separation in spheroidal coordinates which are solutions of some transcendental equation [40], $c = kf$, and $2f$ is the interfocus distance.

The Cartesian coordinates x, y, and z and the spheroidal coordinates ξ and η are related as $x = f\left[(\xi^2 - 1)(1 - \eta^2) \right]^{1/2} \cos\varphi$, $y = f\left[(\xi^2 - 1) (1 - \eta^2) \right]^{1/2} \sin\varphi$, and $z = f\xi\eta$, $1 \leq \xi < \infty$, $-1 \leq \eta \leq 1$, $0 \leq \varphi \leq 2\pi$.

The surface of the spheroid is given by the equation $\xi = \xi^0$, and

$$
ds = f^2 \left[(\xi^{02} - 1)(\xi^{02} - \eta^2) \right]^{1/2} d\eta d\varphi, \quad \frac{\partial}{\partial n} = \sqrt{\frac{\xi^{02} - 1}{\xi^{02} - \eta^2}} \frac{\partial}{\partial \xi}.
$$

We seek the solution to Equation (4.11) in the form

$$
U^\delta(\vec{r}') \big|_S = \sum_{q=0}^{\infty} \sum_{p=-q}^{q} \alpha_{qp}^\delta S_{qp}(c, \eta') e^{ip\varphi'}. \tag{4.26}
$$

Substituting (4.24)–(4.26) into (4.11) and using the orthogonality property of angular spheroidal functions,

$$
\int_{-1}^{1} S_{mn}(c, z) S_{mq}(c, z) dz = N_{mn}(c)\delta_{nq},
$$

we obtain

$$
\alpha_{nm}^\delta = \frac{2(-i)^n \frac{S_{mn}(c, \eta_0)}{N_{mn}(c)} R_{|m|n}^{(1)}(c, \xi_1^0) e^{-im\varphi_0}}{1 + ic(\xi^{02} - 1) R_{|m|n}^{(1)\,\prime}(c, \xi^0) R_{|m|n}^{(4)}(c, \xi_1^0)},
$$

where $\xi = \xi_1^0$ is the equation of the surface S_δ.

Solving the same problem in the exact statement by the method of separation of variables, we obtain

$$\alpha_{nm} = \frac{2(-i)^n}{N_{mn}(c)} \frac{S_{mn}(c, \eta_0) e^{-im\varphi_0}}{ic(\xi^{02} - 1) R^{(4)}_{|m|n}{}'(c, \xi^0)}.$$

Using the relation for the Wronskian [40]

$$R^{(1)}_{mn}(c, \xi) R^{(4)}_{mn}{}'(c, \xi) - R^{(1)}_{mn}{}'(c, \xi) R^{(4)}_{mn}(c, \xi) = \frac{1}{ic(\xi^2 - 1)},$$

we easily obtain the estimate

$$\alpha_{nm} - \alpha_{nm}^\delta \cong \frac{(-i)^n (k\delta)^2}{N_{mn}(c)} \frac{S_{mn}(c, \eta_0) e^{-im\varphi_0}}{R^{(4)}_{|m|n}{}'(c, \xi^0)}$$

$$\times \frac{R^{(1)}_{|m|n}{}'(c, \xi^0) R^{(4)}_{|m|n}{}''(c, \xi^0) - R^{(1)}_{|m|n}{}''(c, \xi^0) R^{(4)}_{|m|n}{}'(c, \xi^0)}{1 + ic(\xi^{02} - 1) R^{(1)}_{|m|n}{}'(c, \xi^0) R^{(4)}_{|m|n}(c, \xi_1^0)}.$$

Thus, we finally have

$$\left| U(\vec{r}')\big|_S - U^\delta(\vec{r}')\big|_S \right| \leq \text{const} \frac{(k\delta)^2}{c(\xi^{02} - 1)}.$$

The last inequality shows that the error, which is determined as the absolute value of the difference between the "currents" on the spheroid surface, obtained by the method of separation of variables and by using the CBCM integral equation (4.11), is proportional to the squared "distance" between the spheroid surface and the surface S_δ.

The above estimates of the error show that, passing from exact boundary conditions to approximate boundary conditions, we obtain the discrepancy of the approximate solution from the exact solution of the order of $k\delta$ for the Dirichlet boundary condition and of the order of $(k\delta)^2$ for the Neumann condition.

4.1.6 Algorithms for Numerical Solution of the CBCM Integral Equations

To solve the CBCM integral equations numerically, we use the Krylov-Bogolyubov technique as in the case of NFM. This chapter mostly deals with three-dimensional problems in the case where S is the surface of a body of revolution and the corresponding algorithm is given in Section 3.1.2. For completeness, we here present the general algorithm and the algorithm for bodies which are less symmetric than the bodies of revolution, i.e., for regular prisms.

For brevity, we illustrate all algorithms with an example of Equation (4.6), i.e., of a boundary condition of the first type.

4.1.6.1 Algorithm for Arbitrary Bodies

1. In the spherical coordinates, $r' = \rho(\theta', \varphi')$ is the equation of the surface S, $r = r(\theta, \varphi)$ is the equation of the surface S_δ, $ds' = \kappa(\theta', \varphi')$ $\rho(\theta', \varphi')d\theta'd\varphi'$, and $\kappa(\theta', \varphi') = \sqrt{(r^2 + \rho_{\theta'}'^2)\sin^2\theta' + \rho_{\varphi'}'^2}$. We denote $\frac{\partial U(\vec{r}')}{\partial n'}\big|_s = \frac{1}{\kappa\rho}I(\theta', \varphi')$. Then integral equation (4.6) in spherical coordinates becomes

$$\int_0^{2\pi}\int_0^\pi I(\theta', \varphi')G_0(\theta', \varphi', \theta, \varphi)d\theta'd\varphi' = U^0(\theta, \varphi, r(\theta, \varphi)), \qquad (4.27)$$

where $G_0 = \frac{1}{4\pi}\cdot\frac{e^{-ikR(\theta', \varphi', \theta, \varphi)}}{R(\theta', \varphi', \theta, \varphi)}$, $\cos\gamma = \sin\theta\sin\theta'\cos(\varphi - \varphi') + \cos\theta\cos\theta'$,

and $R(\theta', \varphi', \theta, \varphi) = \sqrt{\rho^2(\theta', \varphi') + r^2(\theta, \varphi) - 2\rho(\theta', \varphi')r(\theta, \varphi)\cos\gamma}$.

2. We use the projection method to solve Equation (4.27). We piecewise constantly approximate the unknown function in both variables:

$$I(\theta', \varphi') = \sum_{n=1}^N\sum_{m=1}^M I_{n,m}\cdot\phi_{nm}(\theta', \varphi'), \qquad (4.28)$$

where $\phi_{nm}(\theta', \varphi') = \begin{cases} 1, \theta' \in [(n-1)\pi/N, n\pi/N)), & \varphi' \in [(m-1)2\pi/M, m2\pi/M), \\ 0 \text{ otherwise}, \end{cases}$

and substitute representation (4.28) into integral equation (4.27).

3. We divide the variation interval $\theta \in [0, \pi]$ into N equal parts and the variation interval $\varphi \in [0, 2\pi]$ into M equal parts and write the equations at the collocation points coinciding with the middle points of the variation intervals:

$$\theta_q = \frac{\pi}{N}(q-0.5), \quad q = \overline{1,N}, \quad \varphi_h = \frac{2\pi}{M}(h-0.5), \quad h = \overline{1,M}$$

As a result, we obtain the following $NM \times NM$ system of linear algebraic equations (SLAE) for the coefficients of the current expansion in a piecewise constant basis:

$$\sum_{n,m=1}^{N,M} I_{nm}\int_{(m-1)2\pi/N}^{m2\pi/M}\int_{(n-1)\pi/N}^{n\pi/N} G_0(\theta', \varphi', \theta_q, \varphi_h)d\theta'd\varphi' = U^0(\theta_q, \varphi_h, r(\theta_q, \varphi_h)).$$

$$(4.29)$$

4. Further, we can solve SLAE (4.29) directly or pass to discrete sources

$$\frac{2\pi^2}{NM}\sum_{n,m=1}^{N,M} I_{nm}G_0\left(\theta_n,\varphi_m,\theta_q,\varphi_h\right) = U^0\left(\theta_q,\varphi_h,r\left(\theta_q,\varphi_h\right)\right), \qquad (4.30)$$

where $\theta_n=\frac{\pi}{N}(n-0.5)$, $n=\overline{1,N}$, $\varphi_m=\frac{2\pi}{M}(m-0.5)$, $m=\overline{1,M}$.

So the problem of diffraction at an arbitrarily shaped scatterer is reduced to a single SLAE of a very large size $NM\times NM$, which rapidly increases with increasing number of collocation points, and the algorithm converges very slowly in this case. This fact makes the algorithm inefficient if the diffraction problems are solved for bodies whose dimensions are greater than the length of the incident wave. A greater rate of convergence can be obtained if the scatterer is symmetric to some extent.

Let us test this algorithm. We consider the problem of diffraction of a plane wave (3.21) at a three-axial ellipsoid with semiaxes $ka=0.5$, $kb=0.7$, and $kc=1$. In all calculations, $k\delta$ was chosen to be equal to 10^{-3}. Because of the large size of the SLAE obtained for solving this problem numerically, we pass to discrete sources to increase the algorithm realization. Obviously, the solution accuracy and the convergence can be improved if we calculate the corresponding integrals more precisely rather than pass to the sources.

Figure 4.5a and b, respectively, represent the scattering patterns and the boundary condition discrepancy calculated at points between the collocation points for the ellipsoid obtained for $N=50$ and $M=35$.

4.1.6.2 Algorithm for Regular Prisms

Now we use an example of regular prisms to show how the above algorithm can be modified for bodies with some symmetry properties.

1. We consider the problem of diffraction at a regular prism with P faces whose surface is determined by the following equation in rectangular coordinates:

$$(x-a)(x\cos\alpha+y\sin\alpha-a)\ldots(x\cos(P-1)\alpha+y\sin(P-1)\alpha-a)$$
$$\left(z^2-c^2\right)=0,$$

where a is the distance from the prism cross-section center to its face, $2c$ is the prism height, $\alpha=\frac{2\pi}{P}$, $P\geq 3$ is the number of prism faces.

We present the algorithm used to solve this problem. Item 1 remains the same as in the general algorithm for nonaxisymmetric bodies.

2. We perform the change of variables $\varphi'=\psi'+(s-1)\alpha$, $\varphi=\beta+(p-1)\alpha$, $\psi',\beta\in[-\alpha/2,\alpha/2]$, $s=\overline{1,P}$, $p=\overline{1,P}$, and expand the unknown

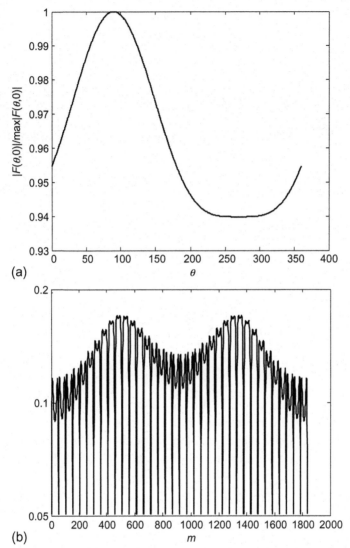

Figure 4.5 (a) Scattering pattern for ellipsoid. (b) Boundary condition discrepancy for ellipsoid.

function, the kernel and the right-hand side of equation (4.27) in the Fourier series in the number of the prism face (which is equivalent to the use of the discrete Fourier transform):

$$I(\theta', \psi' + (s-1)\alpha) = \sum_{l=1}^{P} I^l(\theta', \psi') \cdot e^{i(l-1)(s-1)\alpha}, \tag{4.31}$$

$$G_0(\theta', \psi' + (s-1)\alpha, \theta, \beta + (p-1)\alpha) = G_{sp}(\theta', \psi', \theta, \beta)$$
$$= \sum_{q=1}^{P} G_q \cdot e^{i(q-1)(p-s)}, \qquad (4.32)$$

$$U^0(\theta, \beta + (p-1)\alpha, \rho(\theta, \beta)) = \sum_{\nu=1}^{P} U_\nu^0(\theta, \beta) e^{i(\nu-1)(p-1)\alpha}. \qquad (4.33)$$

3. With $\sum_{l=1}^{P} e^{i(l-1)(p-q)\alpha} = P \cdot \delta_{pq}$ taken into account, we obtain P independent integral equations

$$\int_{-(\alpha/2)}^{\alpha/2} \int_0^\pi I^q(\theta', \psi') G_q(\theta', \psi', \theta, \beta) d\theta' d\psi' = \frac{1}{P} U_q^0(\theta, \beta), \quad q = \overline{1, P}, \quad (4.34)$$

where $\quad G_q(\theta', \psi', \theta, \beta) = \frac{1}{P} \sum_{t=1}^{P} G_{0t}(\theta', \psi', \theta, \beta) e^{i(t-1)(q-1)\alpha} \quad$ and
$U_q^0(\theta, \beta) = \frac{1}{P} \sum_{\nu=1}^{P} U^0(\theta, \beta + (\nu-1)\alpha) e^{i(\nu-1)(q-1)\alpha}$.

4. We piecewise constantly approximate the harmonics of the Fourier expansion of an unknown function in both of the variables:

$$I^q(\theta', \psi') = \sum_{n=1}^{N} \sum_{m=1}^{M_p} I_{n,m}^q \cdot \phi_{nm}(\theta', \psi'), \qquad (4.35)$$

where $\phi_{nm}(\theta', \psi' + (s-1)\alpha) = \begin{cases} 1, & \theta' \in [(n-1)\pi/N, n\pi/N), \\ & \psi' \in [(m-1)\alpha/M_p - \alpha/2, m\alpha/M_p - \alpha/2), \\ 0 & \text{otherwise}, \end{cases}$

and substitute this representation in Equation (4.34).

5. Further, we apply the collocation method to each of the P integral equations (4.34). As a result, we obtain P $NM_p \times NM_p$ SLAEs:

$$\int_{((m-1)\alpha/M_p)-\alpha/2}^{(m\alpha/M_p)-\alpha/2} \int_0^\pi I_{nm}^q(\theta', \psi') G_q(\theta', \psi', \theta_j, \beta_h) d\theta' d\psi' = \frac{1}{P} U_q^0(\theta_j, \beta_h),$$
$$q = \overline{1, P},$$

$$(4.36)$$

where $\theta_j = \frac{\pi}{N}(j - 0.5)$, $j = \overline{1, N}$, $\beta_h = \frac{\alpha}{M_p}(h - 0.5) - \frac{\alpha}{2}$, $h = \overline{1, M_p}$.

6. Then, as previously, we can pass to discrete sources

$$\sum_{n, m=1}^{N, M_p} I_{nm}^q G_q(\theta_n, \psi_m, \theta_j, \beta_h) = \frac{1}{P} U_q^0(\theta_j, \beta_h), \quad q = \overline{1, P}, \qquad (4.37)$$

where

$$\theta_n = \frac{\pi}{N}(n - 0.5), \quad n = \overline{1, N}, \quad \psi_m = \frac{\alpha}{M_p}(m - 0.5) - \frac{\alpha}{2}, \quad m = \overline{1, M_p}. \quad (4.38)$$

The obtained algorithm allows us to decrease the problem dimension. Using this algorithm, we obtain P SLAEs whose size is P times less than the corresponding size in the algorithm for an arbitrary body, which is more profitable from the computational standpoint. In this case, the greater the number of the prism faces, the lesser the size of each of the obtained SLAEs, because the number of points on each face M_p decreases. Moreover, we can expect that the rate of convergence of such an algorithm is high, and this is confirmed by numerical calculations.

To verify the reliability of the results obtained by this algorithm, we calculated the scattering pattern for a 20-face prism and the circular cylinder which is well approximated by such a prism. The calculations for the circular cylinder were performed by the algorithm described in Section 3.1.2, and the calculations for the prism, by the algorithm for regular prisms. As the initial field, we again took the plane wave (3.23).

Figure 4.6a shows the scattering pattern for the regular prism with $P=20$ faces for $ka=2$ (the distance from the center to the face) and $kc=2$ (half the

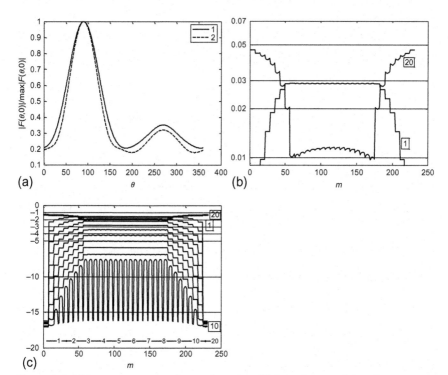

(a)

(b)

(c)

Figure 4.6 (a) Scattering pattern for regular prism. (b) Boundary condition discrepancies for the prism on the first and 20th faces. (c) Boundary condition discrepancies for the prism on all faces where the discrepancies are distinct.

height), and for the cylinder, for $ka=2$ and $kc=2$ (curve 1). The patterns were obtained for $M=6$, $N=N_1=32$, and $Q=4$ ($2Q+1$ is the number of Fourier harmonics in the expansion of the unknown function). One can see that the patterns are sufficiently similar, which testifies that the obtained results are correct. Figure 4.6b and c shows the boundary condition discrepancies for the prism described above (in Fig. 4.6b, on the first and 20th faces, in Fig. 4.6c, on all faces where the discrepancies are distinct). The discrepancies on the faces with numbers from 11 to 19 coincide with the graphs already show in Fig. 4.6c.

4.2 METHOD OF CONTINUED BOUNDARY CONDITIONS FOR VECTOR PROBLEMS OF DIFFRACTION

4.2.1 Statement of the Problem and Derivation of the CBCM Integral Equation

We consider the problem of diffraction of an arbitrary wave field $\vec{E}^0, \vec{H}^0$ at a compact scatterer occupying the spatial domain $\bar{D}$ bounded by the surface S. This means that, in the domain $D_e = \mathbb{R}^3 \backslash \bar{D}$, it is required to find the secondary (diffraction) field $\vec{E}^1, \vec{H}^1$ which satisfies system (3.17) of homogeneous Maxwell equations everywhere outside S and, on the surface S, a boundary condition, for example, of the form

$$\alpha \left[\vec{n} \times \vec{E} \right]_S + \beta \left[\vec{n} \times \left(\vec{n} \times \vec{H} \right) \right]_S = 0, \tag{4.39}$$

where $\vec{E} = \vec{E}^0 + \vec{E}^1$, $\vec{H} = \vec{H}^0 + \vec{H}^1$ is the complete electromagnetic field; the field must also satisfy the Sommerfeld radiation condition at infinity

$$\left(\vec{E}^1 \times \frac{\vec{r}}{r} \right) + \zeta \vec{H}^1 = o\left(\frac{1}{r}\right), \quad \left(\vec{H}^1 \times \frac{\vec{r}}{r} \right) - \frac{1}{\zeta} \vec{E}^1 = o\left(\frac{1}{r}\right), \quad r \to \infty. \tag{4.40}$$

As in the case of scalar problem, the case $\alpha = 1, \beta = 0$ in the boundary condition (4.39) corresponds to the ideal conductor, the case $\alpha = 0, \beta = 1$ corresponds to the ideal magnetic conductor, and the case $-\dfrac{\beta}{\alpha} = Z$ (Z is the surface impedance) corresponds to the impedance scatterer.

According to CBCM, we replace the boundary condition (4.39) on the surface S by an approximate boundary condition on the surface S_δ located at a certain sufficiently small distance δ from S:

$$\alpha \left[\vec{n} \times \vec{E} \right]_{S_\delta} + \beta \left[\vec{n} \times \left(\vec{n} \times \vec{H} \right) \right]_{S_\delta} = 0. \tag{4.41}$$

Now we can derive the CBCM integral equations for different types of boundary conditions. The Maxwell equations (3.17) imply the following integral representations of the secondary field which hold everywhere outside S [31] (also see formulas (1.25)):

$$\vec{E}^1 = \int_S \left\{ \frac{\zeta}{ik} \left(\nabla \times \left(\nabla \times \vec{J}^e G_0(\vec{r},\vec{r}') \right) \right) - \left(\nabla \times \vec{J}^m G_0 \right) \right\} ds', \quad (4.42)$$

$$\vec{H}^1 = \int_S \left\{ \frac{1}{ik\zeta} \left(\nabla \times \left(\nabla \times \vec{J}^m G_0(\vec{r},\vec{r}') \right) \right) + \left(\nabla \times \vec{J}^e G_0 \right) \right\} ds', \quad (4.43)$$

where $\vec{J}^e = \left[\vec{n}' \times \vec{H} \right]_S = \vec{J}$ and $\vec{J}^m = \left[\vec{E} \times \vec{n}' \right]_S$ are the electric and magnetic currents, respectively. To obtain the integral equations, it suffices to have one of these representations. It is expedient to choose the representation that permits obtaining an integral equation with a simpler kernel. But in the majority of cases (except for the integral equations of the first kind for an ideally conducting scatterer), both of these representations lead to integral equations whose kernels are of approximately the same complexity. Therefore, for definiteness, we always use representation (4.42) except for the case specified above. The scheme for deriving the integral equations of the first and second kind is the same as in the scalar problems. Now we apply this scheme.

First, we assume that the scatterer is ideally conducting, i.e., the following boundary condition is satisfied on S_δ:

$$\left[\vec{n} \times \vec{E}^1 \right]_{S_\delta} = - \left[\vec{n} \times \vec{E}^0 \right]_{S_\delta}. \quad (4.44)$$

To obtain the integral equation of the first kind, we use representation (4.42) which, with (4.44) taken into account, can be rewritten as

$$\vec{E}^1 = \frac{\zeta}{i} \int_S \left\{ \frac{1}{k} \left(\vec{J}(\vec{r}') \nabla' \right) \nabla' G_0(\vec{r},\vec{r}') + k \vec{J}(\vec{r}') G_0(\vec{r},\vec{r}') \right\} ds'. \quad (4.45)$$

We substitute representation (4.45) into the boundary condition (4.44) to obtain the Fredholm integral equation of the first kind with a smooth kernel:

$$\frac{\zeta}{i} \int_S \left\{ \vec{n} \times \left[\frac{1}{k} \left(\vec{J}(\vec{r}') \nabla' \right) \nabla' G_0(\vec{r},\vec{r}') + k \vec{J}(\vec{r}') G_0(\vec{r},\vec{r}') \right] \right\} ds' = \vec{n} \times \vec{E}^0,$$

$$(4.46)$$

here and hereafter, $\vec{r}$, $\vec{r}\,'$ are radius vectors of the observation points $M(\vec{r}) \in S_\delta$ and the integration points $N(\vec{r}\,') \in S$, respectively.

To obtain the integral equations of the second kind, we substitute representation (4.43) which, with (4.44) taken into account, can be rewritten as

$$\vec{H}^1 = \int_S \left\{ \vec{J} \times \nabla' G_0 \right\} ds', \tag{4.47}$$

into the identity $\vec{H} = \vec{H}^0 + \vec{H}^1$ and obtain

$$\vec{J}_\delta = \vec{J}^0_\delta + \left[\vec{n} \times \int_S \left\{ \vec{J} \times \nabla' G_0 \right\} ds' \right], \tag{4.48}$$

where $\vec{J}_\delta = \left(\vec{n} \times \vec{H} \right)\big|_{S_\delta}$ and $\vec{J}^0_\delta = \left(\vec{n} \times \vec{H}^0 \right)\big|_{S_\delta}$. Assuming that the surfaces S and S_δ are defined in spherical coordinates by the respective equations $r' = \rho(\alpha, \beta)$ and $r = r(\theta, \varphi)$ and that the quantities $\vec{J}$ and $\vec{J}_\delta$ at the points $\theta = \alpha$, $\varphi = \beta$ are approximately equal to each other, we can consider relation (4.48) as a Fredholm integral equation of the second kind for $\vec{J} \cong \vec{J}_\delta$.

Now we assume that the scatterer is an impedance body (for example, a conducting body with a magnetodielectric coating) [112,114], i.e., the following boundary condition is satisfied on S_δ:

$$\left[\vec{n} \times \vec{E} \right]_{S_\delta} = Z \left[\vec{n} \times \left(\vec{n} \times \vec{H} \right) \right]_{S_\delta}. \tag{4.49}$$

With condition (4.49) taken into account, we can rewrite representation (4.43) as

$$\begin{aligned}
\vec{H}^1 &= \int_S \left\{ \frac{iZ}{k\zeta} \left[k^2 \left(\vec{n}' \times \vec{J} \right) G_0 - \left(\left(\vec{n}' \times \vec{J} \right) \cdot \nabla' \right) \nabla G_0 \right] - \vec{J} \times \nabla G_0 \right\} ds' \\
&= \int_S \left\{ \frac{iZ}{k\zeta} \vec{G}(\vec{r}, \vec{r}') - \vec{J} \times \nabla G_0 \right\} ds',
\end{aligned} \tag{4.50}$$

where $\vec{G}\left(\vec{r}, \vec{r}' \right) = k^2 \left(\vec{n}' \times \vec{J} \right) G_0 - \left(\left(\vec{n}' \times \vec{J} \right) \cdot \nabla' \right) \nabla G_0$.

We substitute representation (4.50) into the boundary condition (4.49), and obtain the Fredholm integral equation of the first kind

$$\int_S \left\{ \frac{Z}{k^2} \left(\vec{n} \times \left(\nabla \times \vec{G}(\vec{r},\vec{r}') \right) \right) - \frac{\zeta}{ik} \left(\vec{n} \times \left(\nabla \times \left(\vec{J} \times \nabla G_0 \right) \right) \right) \right\} ds'$$

$$- Z \int_S \left\{ \frac{iZ}{k\zeta} \left(\vec{n} \times \left(\vec{n} \times \vec{G}\left(\vec{r},\vec{r}' \right) \right) \right) - \vec{n} \times \left(\vec{n} \times \left(\vec{J} \times \nabla G_0 \right) \right) \right\} ds' \; \cdot$$

$$= Z \left(\vec{n} \times \left(\vec{n} \times \vec{H}^0 \right) \right) - \left(\vec{n} \times \vec{E}^0 \right)$$

$$\tag{4.51}$$

Substituting representation (4.50) into the identity $\vec{H} = \vec{H}^0 + \vec{H}^1$, we obtain the Fredholm integral equation of the second kind

$$\vec{J}_\delta = \vec{J}_\delta^0 + \left[\vec{n} \times \int_S \left\{ \frac{iZ}{k\zeta} \left[k^2 \left(\vec{n} \times \vec{J} \right) G_0 - \left(\left(\vec{n} \times \vec{J} \right) \cdot \nabla' \right) \nabla G_0 \right] \right. \right.$$

$$\left. \left. - \vec{J} \times \nabla G_0 \right\} ds' \right]. \quad (4.52)$$

One can see that, in contrast to the scalar problem, the integral equations of the second kind have a simpler form for any type of the boundary condition, and therefore, precisely these equations will be used below.

4.2.2 Algorithm for Solving the CBCM Integral Equations Numerically

The algorithms proposed in Sections 3.1.2 and 4.1.6 for scalar problems can also be used to solve vector problems. Therefore, we briefly consider the algorithm for a body of revolution on which the Dirichlet boundary condition is satisfied, i.e., for integral equation (4.48).

1. We pass to spherical coordinates. Then with the rotational symmetry of the scatterer taken into account, the surfaces S and S_δ are defined by the equations $r' = \rho(\alpha)$ and $r = r(\theta)$ and the auxiliary current $\vec{J}$ can be represented as

$$\vec{J} = \vec{i}_\rho \frac{\rho'(\alpha)}{\rho(\alpha)} J_1 + \vec{i}_\alpha J_1 + \vec{i}_\beta J_2. \tag{4.53}$$

We substitute representation (4.53) into integral equation (4.48), perform several transformations, and obtain the following system of two integral equations for determining the unknown quantities J_1 and J_2:

$$
\begin{cases}
J_1(\theta, \varphi) = J_1^0(\theta, \varphi) + \dfrac{\rho_\delta(\theta)}{\kappa_\delta(\theta)} \displaystyle\int_0^{2\pi} \int_0^{\pi} \{ K_{11}(\theta, \varphi; \alpha, \beta) J_1(\alpha, \beta) \\
\qquad\qquad + K_{12}(\theta, \varphi; \alpha, \beta) J_2(\alpha, \beta) \} \kappa(\alpha) \rho(\alpha) \sin\alpha \, d\alpha \, d\beta, \\[2mm]
J_2(\theta, \varphi) = J_2^0(\theta, \varphi) + \dfrac{1}{\kappa_\delta(\theta)} \displaystyle\int_0^{2\pi} \int_0^{\pi} \{ K_{21}(\theta, \varphi; \alpha, \beta) J_1(\alpha, \beta) \\
\qquad\qquad + K_{22}(\theta, \varphi; \alpha, \beta) J_2(\alpha, \beta) \} \kappa(\alpha) \rho(\alpha) \sin\alpha \, d\alpha \, d\beta.
\end{cases}
\tag{4.54}
$$

In these equations, we have

$$
J_1^0(\theta, \varphi) = \frac{r(\theta)}{\kappa_\delta} H_\varphi^0 \bigg|_{S_\delta}, \quad
J_2^0(\theta, \varphi) = \frac{1}{\kappa_\delta} \left[r(\theta) H_\theta^0 + r'(\theta) H_r^0 \right] \bigg|_{S_\delta},
$$

$$
\kappa(\alpha) = \sqrt{\rho^2(\alpha) + \rho'^2(\alpha)}, \quad \kappa_\delta(\theta) = \sqrt{r^2(\theta) + r'^2(\theta)},
$$

$$
K_{11} = \left\{ -\frac{\partial G_0}{\partial r} \left(\frac{\rho'}{\rho} \frac{\partial \cos\gamma}{\partial \theta} + \frac{\partial^2 \cos\gamma}{\partial \theta \partial \alpha} \right) + \frac{1}{r} \frac{\partial G_0}{\partial \theta} \left(\frac{\rho'}{\rho} \cos\gamma + \frac{\partial \cos\gamma}{\partial \alpha} \right) \right\},
$$

$$
K_{12} = \left\{ -\frac{1}{\sin\alpha} \left(\frac{1}{r} \frac{\partial G_0}{\partial \theta} \frac{\partial \cos\gamma}{\partial \beta} - \frac{\partial G_0}{\partial r} \frac{\partial^2 \cos\gamma}{\partial \theta \partial \beta} \right) \right\},
$$

$$
K_{21} = r \left\{ -\frac{1}{\sin\theta} \left[\frac{1}{r} \frac{\partial G_0}{\partial \varphi} \left(\frac{\rho'(\alpha)}{\rho(\alpha)} \cos\gamma + \frac{\partial \cos\gamma}{\partial \alpha} \right) - \frac{\partial G_0}{\partial r} \left(\frac{\rho'}{\rho} \frac{\partial \cos\gamma}{\partial \varphi} + \frac{\partial^2 \cos\gamma}{\partial \varphi \partial \alpha} \right) \right] \right\}
$$
$$
+ r' \left\{ -\frac{1}{r\sin\theta} \left[\frac{\partial G_0}{\partial \theta} \left(\frac{\rho'}{\rho} \frac{\partial \cos\gamma}{\partial \varphi} + \frac{\partial^2 \cos\gamma}{\partial \varphi \partial \alpha} \right) - \frac{\partial G_0}{\partial \varphi} \left(\frac{\rho'}{\rho} \frac{\partial \cos\gamma}{\partial \theta} + \frac{\partial^2 \cos\gamma}{\partial \theta \partial \alpha} \right) \right] \right\},
$$

$$
K_{22} = r \left\{ -\frac{1}{\sin\theta \sin\alpha} \left(\frac{\partial G_0}{\partial r} \frac{\partial^2 \cos\gamma}{\partial \varphi \partial \alpha} - \frac{1}{r} \frac{\partial G_0}{\partial \varphi} \frac{\partial \cos\gamma}{\partial \beta} \right) \right\}
$$
$$
+ r' \left\{ -\frac{1}{r\sin\theta \sin\alpha} \left(\frac{\partial G_0}{\partial \varphi} \frac{\partial^2 \cos\gamma}{\partial \theta \partial \beta} - \frac{\partial G_0}{\partial \theta} \frac{\partial^2 \cos\gamma}{\partial \varphi \partial \beta} \right) \right\}.
$$

2. We expand the unknown functions, the Green function, and the incident wave in the Fourier series in β, $t = \varphi - \beta$, and φ, respectively,

$$
J_q(\alpha, \beta) = \sum_{n=-\infty}^{\infty} J_{qn}(\alpha) e^{in\beta}, \quad q = 1, 2,
\tag{4.55}
$$

$$
G_0(\alpha, \theta, t) = \sum_{s=-\infty}^{\infty} G_s(\alpha, \theta) e^{ist}
\tag{4.56}
$$

and approximate these series by finite sums consisting of $2Q + 1$ elements (here Q is a small number, because these series converge rapidly).

As a result, because of the rotational symmetry, we obtain a $2Q+1$ system of two integral equations of a lesser dimension.

3. Further, we piecewise constantly approximate the functions $J_{qs}(\alpha)$:

$$J_{qs}(\alpha) = \sum_{n=1}^{N} I_{qn,s} \cdot \phi_n(\alpha), \quad q = 1,2, \tag{4.57}$$

where $\phi_n(\alpha) = \begin{cases} 1, \alpha \in [(n-1) \cdot \pi/N, n \cdot \pi/N], \\ 0, \alpha \notin [(n-1) \cdot \pi/N, n \cdot \pi/N]. \end{cases}$

4. We now apply the collocation method. We obtain $2Q+1$ SLAEs of size $2N \times 2N$, where N is the level or approximation of the unknown functions and hence the number of collocation points:

$$\sum_{n=1}^{N} \Bigg\{ J_{1n,s} \left[\frac{1}{2} - \frac{\pi \rho_\delta(\theta)}{\kappa_\delta(\theta)} \int_{a_n}^{b_n} K_{11s}(\alpha, \theta_m)\kappa(\alpha)\rho(\alpha)\sin\alpha d\alpha \right]$$

$$- \frac{J_{2n,s} \cdot \pi \rho_\delta(\theta)}{\kappa_\delta(\theta)} \int_{a_n}^{b_n} K_{12s}(\alpha, \theta_m)\kappa(\alpha)\rho(\alpha)\sin\alpha d\alpha \Bigg\} = J_{1s}^0(\theta_m) \tag{4.58}$$

$$\sum_{n=1}^{N} \Bigg\{ -\frac{\pi J_{1n,s}}{\kappa_\delta(\theta)} \int_{a_n}^{b_n} K_{21s}(\alpha, \theta_m)\kappa(\alpha)\rho(\alpha)\sin\alpha d\alpha$$

$$+ J_{2n,s} \left[\frac{1}{2} - \frac{\pi}{\kappa_\delta(\theta)} \int_{a_n}^{b_n} K_{22s}(\alpha, \theta_m)\kappa(\alpha)\rho(\alpha)\sin\alpha d\alpha \right] \Bigg\} = J_{2s}^0(\theta_m) \tag{4.59}$$

where $\theta_m = \frac{(m-(1/2))\pi}{N}$, $m = \overline{1, N}$, $a_n = \frac{(n-1)\pi}{N}$, $b_n = \frac{n\pi}{N}$, $s = \overline{-Q, Q}$ and

$$K_{11s}(\alpha, \theta_m) = -k \Big\{ \left(G_{s-1}^r + G_{s+1}^r \right)\cos\theta_m \cdot B_1(\alpha) + 2G_s^r \sin\theta_m \cdot B_3(\alpha)$$

$$- \left(G_{s-1}^\theta + G_{s+1}^\theta \right)\frac{\sin\theta_m}{r} B_1(\alpha) - \frac{2G_s^\theta \cos\theta_m}{r} B_2(\alpha) \Big\},$$

$$K_{12s}(\alpha, \theta_m) = -k \Big\{ \frac{\sin\theta}{ir} \left(G_{s-1}^\theta - G_{s+1}^\theta \right) + i\cos\theta \left(G_{s-1}^r - G_{s+1}^r \right) \Big\},$$

$$K_{21s}(\alpha, \theta_m) = -\frac{k \cdot i}{\sin\theta_m} \Big\{ ((s-1)G_{s-1} + (s+1)G_{s+1})B_1(\alpha)\sin\theta_m$$

$$+ 2s\, G_s B_2(\alpha)\cos\theta_m \Big\} + k \cdot i \cdot B_1(\alpha) \Big\{ r\left(G_{s-1}^r - G_{s+1}^r \right) - \frac{r'}{r}\left[\left(G_{s-1}^\theta - G_{s+1}^\theta \right) \right.$$

$$+ ((s-1)G_{s-1} + (s+1)G_{s+1})\text{ctg}\theta_m - 2s \cdot G_s \cdot \frac{B_3(\alpha)}{B_1(\alpha)} \Big] \Big\},$$

$$K_{22s}(\alpha, \theta_m) = -k \Big\{ r\left(G_{s-1}^r + G_{s+1}^r \right) + ((s-1)G_{s-1}$$

$$- (s+1)G_{s+1}) \cdot \left(\frac{r'}{r}\text{ctg}\theta_m - 1 \right) - \frac{r'}{r}\left(G_{s-1}^\theta + G_{s+1}^\theta \right) \Big\},$$

$$B_1(\alpha) = \left(\frac{\rho'}{\rho}\sin\alpha + \cos\alpha\right), \quad B_2(\alpha) = \left(\frac{\rho'}{\rho}\cos\alpha + \sin\alpha\right), \quad B_3(\alpha)$$

$$= \sin\alpha - \frac{\rho'}{\rho}\cos\alpha$$

here the superscripts on the functions G_s denote the differentiation with respect to the corresponding variables. Because of the double-layer potential jump, the multiplier $1/2$ stands before the unknown function which is not contained in the integrand (see Section 4.3.1).

5. Further, we can pass to discrete sources (see Section 4.3.5):

$$\sum_{n=1}^{N}\left\{J_{1n,s}\left[\frac{1}{2} - \frac{\pi\rho_\delta(\theta)}{4N\kappa_\delta(\theta)}K_{11s}(\alpha_n, \theta_m)\kappa(\alpha_n)\rho(\alpha_n)\sin\alpha_n\right]\right.$$
$$\left. - \frac{J_{2n,s}\cdot\pi\rho_\delta(\theta)}{4N\kappa_\delta(\theta)}K_{12s}(\alpha_n, \theta_m)\kappa(\alpha_n)\rho(\alpha_n)\sin\alpha_n\right\} = J_{1s}^0(\theta_m) \tag{4.60}$$

$$\sum_{n=1}^{N}\left\{-\frac{J_{1n,s}\cdot\pi}{4N\kappa_\delta(\theta)}K_{21s}(\alpha_n, \theta_m)\kappa(\alpha_n)\rho(\alpha_n)\sin\alpha_n\right.$$
$$\left. + J_{2n,s}\left[\frac{1}{2} - \frac{\pi}{4N\kappa_\delta(\theta)}K_{22s}(\alpha_n, \theta_m)\kappa(\alpha_n)\rho(\alpha_n)\sin\alpha_n\right]\right\} = J_{2s}^0(\theta_m) \tag{4.61}$$

where $\alpha_n = \frac{(n-0.5)\pi}{N}$.

4.3 RESULTS OF NUMERICAL INVESTIGATIONS

In this section, we consider different aspects of the CBCM algorithm implementation and determine the optimal parameters of modeling. The CBCM efficiency is estimated by the following characteristics: accuracy, convergence, and reliability of the obtained results. We also discuss the results of the method applications to various problems of diffraction at compact scatterers with constant and variable impedance and at thin screens.

As previously, to estimate the error of a numerical solution, we calculate the discrepancy of the boundary condition (3.22). To estimate the internal convergence of the computational algorithm, we study the rate of decrease in the discrepancy as the approximation parameter N increases and the rate of increase in the accuracy of the optical theorem (3.24) (the rate of decrease in Δ_1).

Unless otherwise specified, all calculations in this section are performed for bodies of revolution by using the algorithm discussed in Section 3.1.2 with the transition to discrete sources. Everywhere except for Section 4.3.3, the surface S_δ is constructed by increasing the radius vector of the surface S by $k\delta$. Unless otherwise specified, the plane wave (3.23) is considered as the initial field.

4.3.1 Taking the Double Layer Potential Jump into Account

The CBCM integral equations of the first kind (4.6)–(4.8) can be solved just as in MAC, i.e., either by using the method of discrete sources or by using a more precise (for example, by splines) approximation of the sought current. But in contrast to MAC, the boundary-value problem in CBCM can also be reduced to a Fredholm integral equation of the second kind (see (4.10)–(4.12)). The procedure of solving these equations is rather special [87].

The CBCM integral equations of the second kind (4.10)–(4.12) differ from the classical current integral equation by the fact that there is no multiplier $1/2$ before the unknown function, which does not enter the integrand; this factor appears due to the double layer potential jump on the scatterer surface. Obviously, such a jump cannot occur at a distance δ from the scatter on which the boundary condition is posed.

But when Equations (4.10)–(4.12) are solved numerically, it is necessary to add this factor, because, as a result of approximate calculations of the coefficients of the Fourier expansion of the kernel and of the unknown function approximation with respect to the angle θ, the space θ, φ is "discretized." In CBCM, it is required that the value δ be sufficiently small, and, therefore, it can be much less than the discretization step both in φ and in θ. Thus, we can assume that the surfaces S and S_δ coincide in the "discrete" space (because the points of these surfaces belong to the same site of the partition for the same values of θ and φ). We also note that the CBCM statement is, therefore, exact in the framework of the numerical implementation (see Section 4.3.2). Thus, there is a jump in the potential, and the multiplier $1/2$ appears in the final SLAE. For example, in the case of boundary condition of the third type, the SLAE has the form

$$\sum_{n=1}^{N} I_{n,s}\left(\frac{1}{2}\cdot\frac{Z}{ik\zeta} + 2\pi\int_{(n-1)\pi/N}^{n\pi/N} K_s(\theta_m, \theta')d\theta'\right) = U_s^0(\theta_m), \qquad (4.62)$$

where $s = \overline{-Q, Q}$, $m = \overline{1, N}$,

$$K_s(\theta, \theta') = G_s(\theta, \theta')\left\{1 - \frac{Z}{ik\zeta R(\theta, \theta', t_\nu)}\left(ik + \frac{1}{R}\right)\frac{1}{\kappa(\theta')}\left(r^2 - r\rho\cos\gamma(\theta, \theta', t_\nu) + r'\rho\frac{\partial\cos\gamma(\theta, \theta', t_\nu)}{\partial\theta'}\right)\right\}.$$

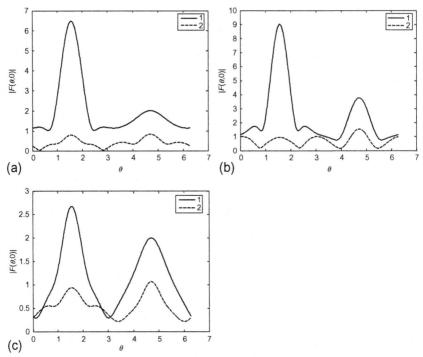

Figure 4.7 (a) Patterns of plane wave scattering at a spheroid with semiaxes $ka=2$ and $kc=4$. (b) Patterns of scattering at a circular cylinder with semiaxes $ka=2$ and $kc=4$. (c) Scattering patterns for impedance spheroid with semiaxes $ka=2$ and $kc=4$.

The fact that the double-layer potential jump must be taken into account is also confirmed by calculations. Figure 4.7a shows the patterns of the plane wave (3.23) scattering at a spheroid with semiaxes $ka=2$ and $kc=4$ (the major semi-axis is directed along the axis z) on which boundary conditions of the first type are posed with the jump taken into account (curve 1) and without the jump (curve 2). The patterns are obtained by solving integral equation (4.10) numerically with the following parameters: $N=N_1=100$, $Q=4$, $k\delta=10^{-5}$ (Q is the maximal number of the azimuthal harmonics). Pattern 1 graphically coincides with the pattern obtained by solving integral equation (4.6) of the first type numerically and with the pattern given in Ref. [42], pattern 2 is significantly different and false.

A similar picture holds for other geometries. For example, Fig. 4.7b shows the patterns of scattering at a circular cylinder with semiaxes $ka=2$ and $kc=4$ (a is the radius, c is the cylinder half-height) on which a boundary condition of the first type is also satisfied. Here the notation and the computational parameters are the same as in the preceding case.

The situation is the same when integral equation (4.12) is solved numerically. Figure 4.7c shows the scattering patterns for a spheroid with semiaxes $ka = 2$ and $kc = 4$ on which a boundary condition of the third type is satisfied for $W = 1000$. The notation is the same as above, the computational parameters are: $N = 250$, $N_1 = 100$, $Q = 4$, and $k\delta = 10^{-5}$.

If the parameter $k\delta$ is greater than the discretization step, the obtained results are false (see Section 4.3.2).

4.3.2 Determining the Value of the Parameter $k\delta$ of the Boundary Condition Continuation

Obviously, the smoother the integral equation kernel, i.e., the greater the value of $k\delta$, the faster the convergence of the computational algorithm. But in contrast, as shown in Section 4.1.5, the greater $k\delta$ in CBCM, the greater the intrinsic error of the solution.

In addition to the error due to continuation of the boundary conditions, the procedure of numerical solution of the integral equations also gives an error of approximation of the unknown function $I(\theta, \varphi)$, which is determined by the approximation levels N and $Q \ll N$, and an error of approximation of the Fourier harmonics of the integral equation kernel, which is determined by the approximation level $N_1 \leq N$. Therefore, to decrease the solution error due to the displacement of the boundary condition, it is expedient to take the value of $k\delta$ less than the step of the unknown function approximation with respect to θ:

$$|k\delta| < \frac{\pi}{N}. \tag{4.63}$$

Then in the framework of the introduced partition, we can assume that the statement of the problem is exact and can expect to obtain a stable solution. The analyticity of the wave field guarantees that the boundary condition is satisfied approximately in some usually small neighborhood. Therefore, the choice of $k\delta \gg \frac{\pi}{N}$ can lead to false results.

Moreover, if the boundary-value problem is reduced to a Fredholm equation of the second kind, then because it is necessary to take the potential jump into account, the value of $k\delta$ must be less than the approximation step at least by an order of magnitude, i.e., it is necessary to sharpen condition (4.63) for the integral equation of the second kind:

$$|k\delta| \ll \frac{\pi}{N}. \tag{4.64}$$

We use a specific example to show that it is necessary to satisfy conditions (4.63) and (4.64) and use numerical experiments to specify inequality (4.64).

We again consider the problem of diffraction of a plane wave at a spheroid with semiaxes $ka = 2$ and $kc = 4$ on which a boundary condition of the first type is satisfied. First, we assume that the problem is already reduced to the Fredholm integral equation of the first kind (4.6). Figure 4.8a shows the discrepancies of the boundary condition on the surface S_δ calculated for $N = N_1 = 100$ (i.e., condition (4.63) becomes $|k\delta| < 0.03$) and for different values of $k\delta$: curve 1—$k\delta = 10$, 2—$k\delta = 1$, 3—$k\delta = 0.1$, 4—$k\delta = 0.01$, 5—$k\delta = 0.001$, and 6—$k\delta = 0.0001$. As $k\delta$ continues to decrease, the discrepancy practically ceases to vary. One can see that as $k\delta$ decreases, the discrepancies become worse starting from $k\delta = 1$. As previously noted, this is related to the fact that the smoothness of the integral equation kernel decreases. Thus, the best satisfaction of the

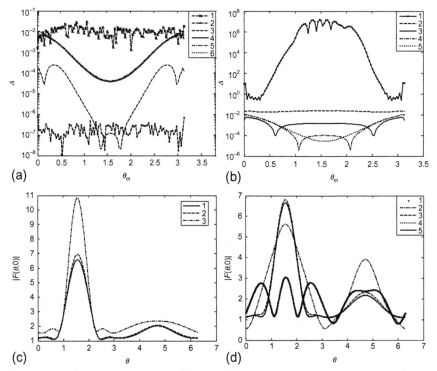

Figure 4.8 (a) Boundary condition discrepancies on the surface S_δ for spheroid with semiaxes $ka=2$ and $kc=4$. (b) Boundary condition discrepancies on the surface S. (c) Scattering patterns for spheroid with semiaxes $ka=2$ and $kc=4$. (d) Scattering patterns for spheroid with semiaxes $ka=2$ and $kc=4$ in the problem reduced to an integral equation of the second kind.

approximate boundary conditions is attained for $k\delta = 1$. But we should not take $k\delta \geq 0.03$, because the exact boundary condition on the surface S is then significantly worse and condition (4.63) is not satisfied.

Figure 4.8b shows the discrepancies of the boundary condition on the surface S calculated for $N = N_1 = 100$ and for different values of $k\delta$: curve 1—$k\delta = 1$, 2—$k\delta = 0.1$, 3—$k\delta = 0.01$, 4—$k\delta = 0.001$, and 5—$k\delta = 0.0001$.

One can see that the accuracy of the boundary condition on the surface S increases as $k\delta$ decreases and hence the solution obtained by CBCM tends to the exact solution as $k\delta \rightarrow 0$. We note that the values of discrepancies near the points $\theta = 0$ and $\theta = \pi$ for $k\delta = 0.001$ and $k\delta = 0.0001$ are somewhat greater than the same values for $k\delta = 0.01$. This can be explained by the fact that decreasing $k\delta$, we approach the singularities of the Green function at $\theta = 0$ and $\theta = \pi$. Therefore, we should not take too small values of $k\delta$.

The fact that it is necessary to satisfy condition (4.63) is confirmed by calculations of the scattering patterns. As expected, the patterns for $k\delta = 10$ and $k\delta = 1$ are false, and the pattern for $k\delta = 0.1$ was calculated with a noticeable error (see Fig. 4.8c). This figure shows the scattering patterns for the considered problem: 1—$k\delta \leq 0.01$, 2—$k\delta = 0.1$, and 3—$k\delta = 1$.

Now we assume that the considered boundary-value problem is reduced to an integral equation of the second kind. Figure 4.8d shows the patterns of scattering at the same spheroid for different $k\delta$: 1—$k\delta = 0.01$, 2—$k\delta = 10^{-3}$, 3—$k\delta = 10^{-4}$, 4—$k\delta = 10^{-5}$, and 5—$k\delta = 10^{-6}$. The further decrease in $k\delta$ does not improve the result.

Although inequality (4.63) is here satisfied for all curves, the obtained result is false for $k\delta = 0.01$ and $k\delta = 10^{-3}$, and the scattering pattern is calculated with a noticeable error for $k\delta = 10^{-4}$. This result allows us to specify inequality (4.64) as follows:

$$|k\delta| \leq \frac{\pi}{10^3 N} \tag{4.65}$$

Thus, the value of the parameter $k\delta$ of the boundary condition continuation is determined by the size N of SLAE obtained when solving the problem numerically, and this size in turn depends on the relative size of the scatterer.

4.3.3 Choice of the Method for Constructing the Surface Where the Boundary Condition Is Satisfied

It is necessary to choose a surface S_δ so that all its points be from the surface S at a distance that does not exceed $k\delta$. Obviously, such surfaces can be constructed in many different ways. For example, S_δ can be constructed so that all its points are at a distance exactly equal to $k\delta$ from the surface S. Such a surface is said to be equidistant to S. For analytic surfaces, an equidistance surface (located at a sufficiently small distance from the initial one) can approximately be constructed by using the technique of analytic deformation (see Section 3.6). But in the general case, construction of an equidistant surface requires additional calculations. The surface S_δ can also be constructed by the following different methods, namely, all parameters of the surface are increased by $k\delta$ (for example, the semiaxes of a spheroid or a cylinder) or are multiplied by a certain coefficient $\lambda > 1, \lambda = 1 + \alpha \cdot k\delta$, or the radius vector of the surface S is decreased by $k\delta$. Let us consider how the method used to construct the surface S_δ can affect the accuracy of computations.

Figure 4.9 shows the discrepancies of the boundary condition obtained by solving the problem of diffraction at a spheroid with semiaxes $ka = 2$ and $kc = 4$ on which a boundary condition of the first type is satisfied for $k\delta = 0.01$ and $N = N_1 = 100$. These discrepancies are obtained by different methods for constructing the surface S_δ: curve 1 corresponds to the analytic deformation of S, curve 2 corresponds to the increase in the spheroid semiaxes by $k\delta$, curve 3 corresponds to multiplication of the spheroid semiaxes by the coefficient $\lambda = 1.0025$, and curve 4 corresponds to the increase in the radius vector of the surface S by $k\delta$.

One can see that the difference between the discrepancies is small, and this difference only decreases with decreasing $k\delta$. Therefore, for $k\delta$ satisfying

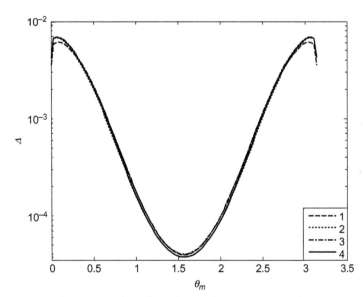

Figure 4.9 Boundary condition discrepancies for spheroid with semiaxes $ka = 2$ and $kc = 4$.

(4.63), the method used to construct the surface S_δ is not important. Thus, the surface S_δ can always be constructed according to the same scheme, namely, by the simplest and universal method of all the methods described in this section: by increasing the radius vector of the surface S by $k\delta$.

4.3.4 Use of Fredholm Integral Equations of the First and Second Kind

As previously noted, the boundary-value problem (4.1), (4.3) can be reduced to integral equations of either the first or the second kind, and this raises the question of choosing the most preferable equation. To answer this question, we compare the efficiency of the computational algorithms based on these equations. We preliminary assume that it is expedient to choose the integral equation whose kernel has a simpler form for the considered boundary condition.

Figure 4.10a shows the patterns of a plane wave scattering at a spheroid with semiaxes $ka = 2$ and $kc = 4$ on which the boundary condition of the first

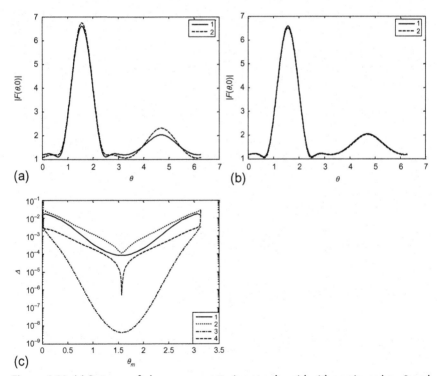

Figure 4.10 (a) Patterns of plane wave scattering at spheroid with semiaxes $ka = 2$ and $kc = 4$, $N = N_1 = 50$. (b) Patterns of plane wave scattering at spheroid with semiaxes $ka = 2$ and $kc = 4$, $N = 350$, $N_1 = 150$. (c) Boundary condition discrepancies.

type is satisfied; the scattering patterns were obtained for $N = N_1 = 50$ and $k\delta = 10^{-5}$: curve 1 was obtained by solving the integral equation of the first kind (4.6), and curve 2, by solving the integral equation of the second kind (4.9). One can see that the pattern based on the integral equation of the second kind is calculated with a noticeable error. The results in Fig. 4.10b were obtained for $N = 350$ and $N_1 = 150$.

The differences in the patterns for the integral equation of the first kind calculated for different N become noticeable "by sight" starting approximately from $N = 50$, while the patterns for the integral equations of the second kind become indistinguishable by sight starting approximately only from $N = 200$. This means that the algorithm based on the integral equation of the first kind, which has a simpler kernel in this case, demonstrates a higher convergence and is preferable. Moreover, the use of the integral equation of the first kind gives an additional gain in time because a lesser amount of arithmetical operations is required to calculate each SLAE coefficient. The better convergence of the computational algorithm based on the integral equation of the first kind is also confirmed by the rate of decrease in the discrepancy. Figure 4.10c illustrates the discrepancies of the boundary condition for different N for the integral equations of the first and second kind: curves 1 and 3 are obtained on the basis of the integral equation of the first kind for $N = N_1 = 50$ and $N = 350$, $N_1 = 150$, respectively, curves 2 and 4, on the basis of the integral equation of the second kind for the same N and N_1, respectively.

The worse convergence of the computational algorithm based on the integral equation of the second kind can be explained by the fact that its kernel contains the derivative of the Green function and hence is a more rapidly varying function near the singularities than the kernel of the integral equation of the first kind. Therefore, a greater amount of nodes is required for a good approximation of the Fourier harmonics of this function. Thus, it is more expedient to use integral equations of the first kind in the case of a boundary condition of the first type. By the same reason, in the case of boundary conditions of the second type, one should use integral equations of the second kind whose kernels, as previously noted, are simpler in form compared with the kernels of integral equations of the first kind.

4.3.5 Transition to Discrete Sources

When implementing the algorithm given in Section 3.1.2, we can stop at stage 5 or can pass to discrete sources following the procedure at stage 6.

The difference is that, in the first case, the integrals must be calculated by any quadrature formula (for brevity, this approach is called the method of currents (MC)) and, in the second case, the integrals are replaced by sums (this approach is called the method of discrete sources (MDS)). Now we consider which of these versions is more efficient.

In all calculations in this section, the integrals are calculated by the Gauss-Kronrod quadrature formula with 21 nodes, which ensures a rather high rate of calculations. We note that, in scalar diffraction problems, the method of numerical integration is not important to such an extent as in vector diffraction problems, where the time of the program implementation is significantly larger.

Figure 4.11a and b illustrates the boundary condition discrepancies obtained for the solution of the diffraction problem at a spheroid and a cylinder with the respective semiaxes (radius and half-height in the case of cylinder) $ka = 2$, $kc = 4$ on which the boundary condition of the first type is satisfied for $k\delta = 10^{-3}$ and $Q = 4$: 1—$N = 32$ (MDS), 2—$N = 128$ (MDS), 3—$N = 16$ (MC), and 4—$N = 32$ (MC); the SLAE coefficients were calculated by the two methods described above. One can see that the results obtained by the method of currents is much more precise and its use allows one significantly to decrease the SLAE size; for example, the method of currents allows one to obtain a greater accuracy for $N = 16$ than MDS ensures for $N = 128$, while the time of the algorithm implementation is approximately the same (68 s for MDS and 60 s for MC). In Fig. 4.11b, curves 2 and 3 differ insignificantly, i.e., the gain obtained for the cylinder by using precisely the method of currents for $N = 16$ is significantly less than that for the spheroid. But if the accuracy attained for $N = 32$ in the case of MDS (for example, for small dimensions of the scatterer) is sufficient, then the use of this method is more profitable, because the program realizing MC for $N = 16$ operates approximately four times longer than the program realizing MDS for $N = 32$.

We note that the discrepancy overshoots near the singularities of the Green function (for $\theta = 0$, $\theta = \pi$) arising in the method of currents are greater than those arising in MDS, which is obvious, because we approach these singularities much closer in MC. The discrepancies in the case of cylinder have overshoots near the break points of the cylinder surface, and these overshoots do not decrease even in MC.

Thus, it is expedient to use MC only if the SLAE size in MDS becomes too large, for example, in the case where a very high accuracy of calculations is required or the scatterer size is much greater than the wavelength.

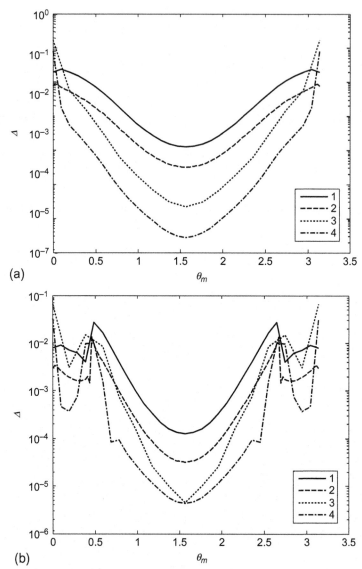

(a)

(b)

Figure 4.11 (a) Boundary condition discrepancies obtained for spheroid. (b) Boundary condition discrepancies obtained for cylinder.

In the algorithm in Section 3.1.2, in addition to the parameter N (the number of discrete sources), there is also a parameter N_1, which is the number of points required to calculate the S-functions by the rectangular formula. At first sight, it seems that these parameters are independent, but in practice, when an algorithm with transition to discrete sources is used, it is very important to choose a correct relation between these parameters. In all calculations in scalar problems, above and below, the value of N is chosen to be equal to N_1. Moreover, the choice $N_1 > N$ leads to false results. But in vector problems, the results are false even for $N = N_1$. Let us consider why the relation between N and N_1 affects the accuracy of calculations.

We note that the vector diffraction problems are more sensitive to the choice of numerical methods because the corresponding integral equations contain the second derivatives of the Green functions. The point is that the integral equations in scalar problems contain only the Green function and its first derivative, which, after the boundary conditions are continued and in the case of coinciding arguments, leads to a quasisingularity of the form

$$\frac{1}{R^2(\vec{r},\vec{r'})} \sim \frac{1}{(k\delta)^2},$$

which does not strongly affect the accuracy of calculations, of course, for a reasonable value of $k\delta$ (of the order of $10^{-3} \div 10^{-6}$). But in vector problems, because of the appearance of the above-mentioned second derivatives, there is already a quasisingularity of the form

$$\frac{1}{R^5(\vec{r},\vec{r'})} \sim \frac{1}{(k\delta)^5},$$

which can lead to a large computational error for $k\delta \leq 10^{-3}$.

Thus, the accuracy of calculations (especially in vector problems) can decrease if $R \approx k\delta$, i.e., if there is a quasisingularity. Because

$$R = \sqrt{r^2(\theta) + \rho^2(\alpha) - 2r(\theta)\rho(\alpha)\cos\gamma} \text{ and } \cos\gamma = \sin\theta\sin\alpha\cos t + \cos\theta\cos\alpha,$$

we have $R \approx k\delta$ if and only if

$$\begin{cases} \theta = \alpha, \\ \cos\gamma = 1, \end{cases}$$

and therefore,

$$\cos\gamma = \sin^2\theta\cos t + \cos^2\theta = 1. \tag{4.66}$$

This equality is satisfied for $t = 0$, $t = 2\pi$ and/or for $\theta = 0$, $\theta = \pi$. These values are never attained in discretization, but the extreme points of the sequences θ_m and t_ν lie near these values (and the closer, the greater N and N_1):

$$\theta = \theta_1 = \frac{\pi}{2N}, \quad \theta = \theta_N = \pi - \frac{\pi}{2N} \quad \text{and} \quad t = t_1 = \frac{\pi}{N_1}, \quad t = t_{N_1} = 2\pi - \frac{\pi}{N_1}$$

$$(4.67)$$

It follows from (4.66) and (4.67) that for the extreme points of the discrete sequences θ_m and t_ν to be as far as possible from the ends of the intervals, it is necessary to choose $N \gg N_1$. On the one hand, the greater the difference between N and N_1, the farther the discrete values from the quasisingularity; on the other hand, a too large distance to the singularity also leads to an error. Thus, the choice of a correct relation between N and N_1 is very important in vector problems.

We illustrate this with an example of the solution of the diffraction problem of a plane electromagnetic wave incident at the angles $\varphi_0 = 0$, $\theta_0 = \pi/2$ at an ideally conducting spheroid with semiaxes $ka = 1$ and $kc = 2$. The calculations were performed for different relations between N and N_1. For $N \leq N_1$, the obtained results are false, and it does not make any sense to present them here. We present the results of calculations obtained for $N > N_1$, $k\delta = 10^{-3}$, and $Q = 4$.

Figure 4.12a shows the scattering patterns in the plane $\varphi = [0, \pi]$ for F_θ^E and in the plane $\varphi = [\pi/2, 3\pi/2]$ for F_φ^E for the above-described spheroid for $N = 64$, $N_1 = 16$ (curves 1) and for $N = 64$, $N_1 = 32$ (curves 2). Figure 4.12b presents the corresponding discrepancies of the boundary condition calculated at points between the collocation points in the plane $\varphi = [0, \pi]$, where they are the worst (dotted curves—$N_1 = 16$, solid curve—$N_1 = 32$). For clarity, the discrepancies are shown only for system (4.59); they are of the order of 10^{-12} for system (4.58), which follows from Fig. 4.14.

Curves 1 in Fig. 4.12a graphically coincide with the patterns obtained by different methods. Thus, despite the fact that twice as many discrete values t_ν were taken to calculate curves 2, they still strongly differ from the correct result. This also follows from Fig. 4.12b, namely, the discrepancy for $N_1 = 16$ is much better than that for $N_1 = 32$. This means that, to obtain a result with at least a graphical accuracy, the relation between N and N_1 must at least be $N \approx 4N_1$, and this also confirms that the above reasoning is correct.

We compare the results obtained by passing to discrete sources with the results obtained without such a transition. Figure 4.13a and b represents the respective scattering patterns in the plane $\varphi = [0, \pi]$ for F_θ^E and in the plane $\varphi = [\pi/2, 3\pi/2]$ for F_φ^E and the discrepancies obtained for $N = 64$ and $N_1 = 16$ (solid curve, with transition to discrete sources, and dotted curve, without transition).

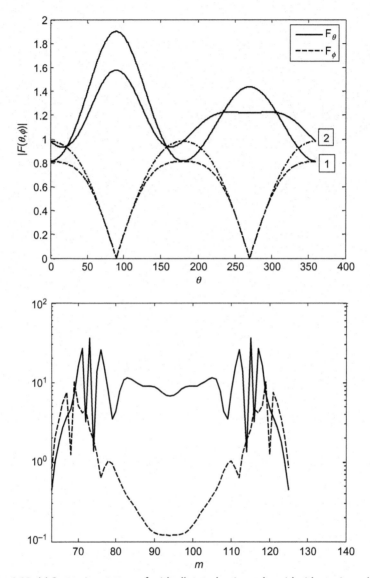

Figure 4.12 (a) Scattering patterns for ideally conducting spheroid with semiaxes $ka = 1$ and $kc = 2$. (b) Boundary condition discrepancies.

One can see that the patterns differ insignificantly, and the discrepancy becomes worse after the transition to discrete sources. But the computation time is six times shorter after the transition to discrete sources! This allows us to obtain a greater accuracy by increasing N.

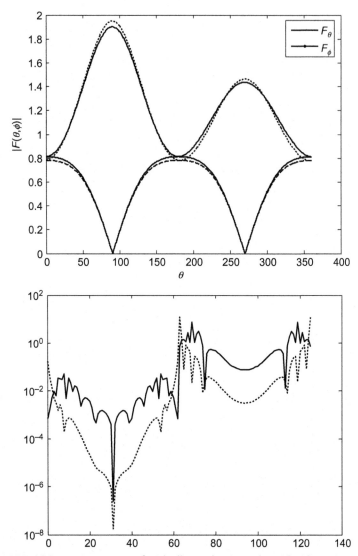

Figure 4.13 (a) Scattering patterns for ideally conducting spheroid with semiaxes $ka = 1$ and $kc = 2$ with and without transition to discrete sources. (b) Boundary condition discrepancies for ideally conducting spheroid with semiaxes $ka = 1$ and $kc = 2$ with and without transition to discrete sources.

We again try to increase the ratio of N to N_1 by increasing N. Figure 4.14 shows the discrepancies obtained for $N_1 = 16$ and $N = 64$ (dash-dotted curve), $N = 128$ (dotted curve), and $N = 256$ (solid curve). One can see that the algorithm with transition to discrete source allowed us to obtain a much

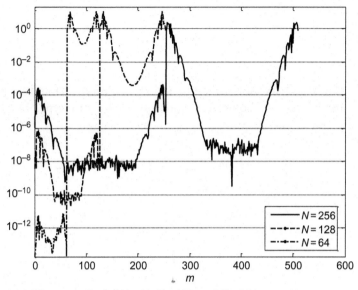

Figure 4.14 Discrepancies for $N = 64$, $N = 128$, and $N = 256$.

greater accuracy in a lesser time than the algorithm without such a transition in calculations for $N = 64$ and $N_1 = 16$.

4.3.6 Choice of a Basis for Approximating the Unknown Function

To improve the accuracy with an insignificant increase in the computation time in scalar and vector problems, we can approximate the unknown function (functions) by using different bases instead of a piecewise constant basis.

To approximate the unknown function in the integral equation, it is preferable to use a local approximation, because the latter obviously decreases the interval of integration simultaneously increasing the level of approximation. Therefore, as approximating functions we use splines of different orders to solve the posed problem. The splines are widely used because of the fact that, in a sense, they are the smoothest functions in the class of functions taking the prescribed values (in the case of interpolation). In the case of a smooth function, the splines of a degree greater than the first ensure a good approximation not only of the function but also of its derivatives.

The piecewise constant approximation is the approximation by zeroth-order splines. We illustrate the approximation by splines of higher orders (up to three inclusively) with an example of a scalar problem of diffraction at a body of revolution on which the Dirichlet boundary condition is posed.

Let an unknown function be represented as an expansion in a basis:

$$I_s(\theta') = B_\phi \sum_{n=1}^{N} I_{n,s} \cdot \phi_n(\theta'). \qquad (4.68)$$

Here the $\phi_n(\theta')$ are basis functions and B_ϕ is a normalizing multiplier.

The in the case of boundary condition of the first type, SLAE has the form

$$B_\phi \sum_{n=1-p}^{N} I_{n,s} \int_{a_n}^{b_n} \phi_n(\theta') G_s(\theta_m, \theta') d\theta' = U_s^0(\theta_m), \qquad (4.69)$$

where p is degree of a spline.

In the case of a piecewise approximation, we have

$$\phi_n = \phi_{0n}(\theta') = \begin{cases} 1, & \theta' \in [(n-1)\pi/N, n\pi/N), \\ 0, & \theta' \notin [(n-1)\pi/N, n\pi/N). \end{cases} \quad B_\phi = 1, \ p = 0,$$

$$\theta_m = \frac{\pi}{N}(m - 0.5), \quad m = \overline{1, N}.$$

In the case of approximation by linear splines, we have

$$\phi_n = \phi_{1n}(\theta') = \begin{cases} \theta' - \alpha_n, \theta' \in [(n-1)\pi/N, n\pi/N), \\ \alpha_{n+1} - \theta', \theta' \in [n\pi/N, (n+1)\pi/N), \\ 0 \ \text{otherwise}, \end{cases} \quad B_\phi = \frac{N}{\pi}, \ p = 1,$$

$$\alpha_n = \frac{(n-1)\pi}{N}, \quad n = \overline{1, N-1}, \quad \theta_m = \frac{m\pi}{N}, \quad m = \overline{0, N}$$

and the splines at the ends are "cut":

$$\phi_0(\theta') = \begin{cases} \alpha_0 - \theta', \theta' \in [0, \pi/N), \\ 0 \ \text{otherwise}. \end{cases} \quad \alpha_0 = \frac{\pi}{N},$$

$$\phi_N(\theta') = \begin{cases} \theta' - \alpha_N, \theta' \in [(N-1)\pi/N, \pi), \\ 0 \ \text{otherwise}. \end{cases} \quad \alpha_N = \frac{(N-1)\pi}{N}.$$

The higher-order splines at the ends are constructed similarly, and we do not present them here.

In the case of approximation by quadratic splines, we have

$$\phi_n = \phi_{2n}(\theta') = \begin{cases} \dfrac{(\theta' - \alpha_n)^2}{2}, \theta' \in [\alpha_n, \alpha_{n+1}], \\[2mm] -(\theta' - \alpha_n)^2 + \dfrac{3\pi}{N}(\theta' - \alpha_n) - \dfrac{3\pi^2}{2N^2}, \theta' \in [\alpha_{n+1}, \alpha_{n+2}], \\[2mm] \dfrac{(\theta' - \alpha_n)^2}{2} - \dfrac{3\pi}{N}(\theta' - \alpha_n) + \dfrac{9\pi^2}{2N^2}, \theta' \in [\alpha_{n+2}, \alpha_{n+3}], \\[2mm] 0 \text{ otherwise.} \end{cases}$$

$$B_\phi = \frac{4N^2}{3\pi^2},$$

$$p = 2, \quad \alpha_n = \frac{(n-1)\pi}{N}, \quad n = \overline{1, N-2}, \quad \theta_m = \frac{(m+0.5)\pi}{N},$$

$$m = \overline{0, N-1}, \quad \theta_{-1} = 0, \quad \theta_N = \pi.$$

And finally, in the case of approximation by cubic splines, we have

$$p = 3, \quad \alpha_n = \frac{(n-1)\pi}{N}, \quad n = \overline{1, N-3}, \quad \theta_m = \frac{(m+0.5)\pi}{N}, \quad m = \overline{0, N-1}.$$

$$\theta_{-2} = 0, \quad \theta_{-1} = \frac{\pi}{4N}, \quad \theta_{N-1} = \frac{\pi(4N-1)}{4N}, \quad \theta_N = \pi, \quad B_\phi = \frac{3N^3}{2\pi^3},$$

$$\phi_{n3}(x) = \begin{cases} \dfrac{(\theta' - \alpha_n)^3}{6}, \; \theta' \in [\alpha_n, \alpha_{n+1}], \\[2mm] -\dfrac{(\theta' - \alpha_n)^3}{2} + \dfrac{2\pi}{N}(\theta' - \alpha_n)^2 - \dfrac{2\pi^2}{N^2}(\theta' - \alpha_n) + \dfrac{2}{3}\dfrac{\pi^3}{N^3}, \; \theta' \in [\alpha_{n+1}, \alpha_{n+2}], \\[2mm] \dfrac{(\theta' - \alpha_n)^3}{2} - \dfrac{4\pi}{N}(\theta' - \alpha_n)^2 + \dfrac{10\pi^2}{N^2}(\theta' - \alpha_n) - \dfrac{22}{3}\dfrac{\pi^3}{N^3}, \; \theta' \in [\alpha_{n+2}, \alpha_{n+3}], \\[2mm] -\dfrac{(\theta' - \alpha_n)^3}{6} + \dfrac{2\pi}{N}(\theta' - \alpha_n)^2 - \dfrac{8\pi^2}{N^2}(\theta' - \alpha_n) + \dfrac{32}{3}\dfrac{\pi^3}{N^3}, \; \theta' \in [\alpha_{n+3}, \alpha_{n+4}], \\[2mm] 0 \text{ otherwise.} \end{cases}$$

Here the choice of collocation points is of great importance.

It practically does not make any sense to pass to discrete sources in the case of spline approximation, because the obtained system obviously slightly differs from the system obtained by using the piecewise constant approximation.

Let us consider how the above-listed bases can be used to solve specific problems of diffraction.

We consider the problem of diffraction of a plane wave incident at the angles $\varphi_0 = 0$ and $\theta_0 = \pi/2$ at a spheroid with semiaxes $ka = 2$ and $kc = 4$ (the major semiaxis is directed along the axis z). We present the calculation

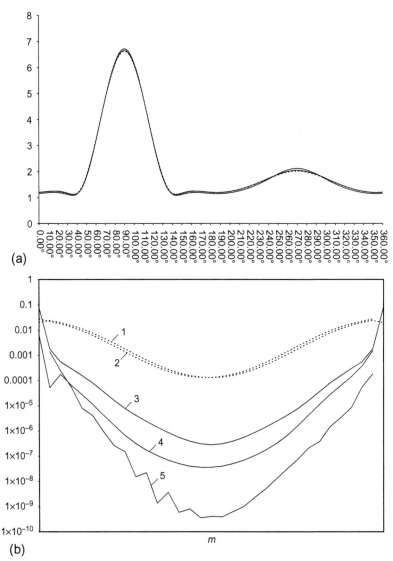

(a)

(b)

Figure 4.15 (a) Scattering patterns for spheroid with semiaxes $ka=2$, $kc=4$ obtained for different bases. (b) Discrepancies for spheroid with semiaxes $ka=2$, $kc=4$ obtained for different bases.

results obtained by using different bases (the spline order varies from zero to two) for $k\delta = 10^{-3}$ and $Q = 4$.

Figure 4.15a shows the scattering patterns obtained by using the above-listed bases for $N = 32$. One can see that they practically coincide, which means that the obtained results are correct.

The corresponding discrepancies of the boundary condition are presented in Fig. 4.15b, where curve 1 corresponds to the piecewise constant approximation with transition to discrete sources (PC-DS below), curve 2 shows the result of approximation by linear splines with transition to discrete sources (LinDS below), curve 3 corresponds to the piecewise constant approximation (PC below), curve 4 shows the result of approximation by linear splines (Lin below), and curve 5 corresponds to approximation by quadratic splines (Quad below).

One can see that in the case of transition to discrete source, the accuracy of the solution obtained by using linear splines is the same as in the case of piecewise constant approximation, i.e., as previously noted, when passing to discrete sources in the case of smooth bodies, it does not make any sense to use splines of order greater than zero. But if we do not pass to discrete sources, then the accuracy of the solution increases with the spline order while the computational time increases only insignificantly.

Now we consider the problem of diffraction of a plane wave incident at the angles $\varphi_0 = 0$ and $\theta_0 = \pi/2$ at a circular cylinder with the base radius $ka = 2$ and half-height $kc = 4$. We present the computation results obtained by using different bases (splines whose order varies from zero to two) for $k\delta = 10^{-3}$ and $Q = 4$. The patterns practically coincide as in the case of spheroid, and hence we do not present them here.

Figure 4.16a shows the discrepancies of the boundary condition for the cylinder: 1—PC-DS, 2—LinDS, 3—PC, 4—Lin, and 5—Quad for $N = 32$.

One can see that in the case of transition to discrete sources, one can avoid the discrepancy overshoots near the break points of the scatterer boundary. But in contrast to the spheroid, the highest accuracy attained in the case of cylinder without passing to discrete sources at a prescribed approximation level is given by the piecewise constant approximation, while the lowest accuracy can be obtained by using quadratic splines.

Let us increase the approximation level. Figure 4.16b illustrates the discrepancies of the boundary condition for the cylinder: 1—LinDS, 2—PC, 3—Lin, and 4—Quad for $N = 128$.

One can see that the picture is somewhat different. Now the best discrepancy can be obtained by using quadratic splines, and the worst discrepancy, by using piecewise constant splines. On the whole, the values of all discrepancies are less compared with the corresponding discrepancies for $N = 32$. But the discrepancy in the case of piecewise constant approximation decreases only by an order of magnitude, while the discrepancy in the case of linear splines decreases almost by three orders, and for the quadratic splines, by five orders! This means that, in the class of considered bases, the highest rate of the algorithm convergence is ensured by quadratic splines, the linear

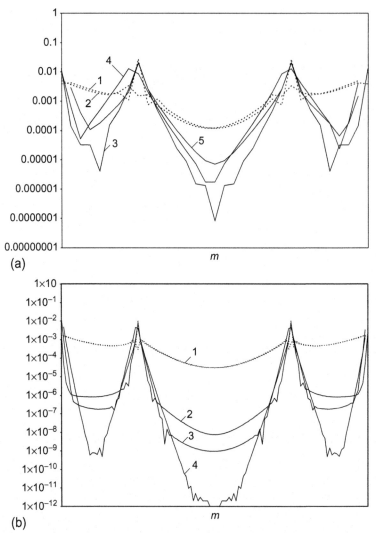

Figure 4.16 (a) Boundary condition discrepancies for cylinder, $N = 32$. (b) Boundary condition discrepancies for cylinder, $N = 128$.

splines give a lesser rate of convergence, and the lowest rate is given by the piecewise constant approximation. Nevertheless, the value of the discrepancy overshoots remains practically unchanged, which can be explained by the fact that the scatterer surface has break points.

Thus, the computational experiments show that the greater the spline order, the higher the rate of the algorithm convergence. But a decrease in the discrepancy as the spline order increases is not manifested immediately but can be seen only for a sufficiently large N, i.e., for a sufficiently large set of basis functions. If N is insufficiently large, then the picture is opposite, i.e., the best result can be obtained by the piecewise constant approximation. Therefore, if a high computational accuracy is not required, then it is expedient to use the piecewise constant approximation. And conversely, if the scatterer dimensions are large and the rate of convergence is of decisive importance, then it is preferable to use splines of a possibly greater order.

4.3.7 Calculation of the Vasiliev S-Functions

The quasisingularities of the kernels of integral equations in vector problems create certain computational difficulties (see Section 4.3.5) and, therefore, it is possibly required to calculate the integrals (Vasiliev S-functions and their first and second derivatives) more precisely near these quasisingularities. Earlier, these integrals were calculated by the rectangular formula, which is equivalent to the transition from the direct continuous Fourier transform (DCFT) to the direct discrete Fourier transform (DDFT), for example,

$$
\begin{aligned}
G_s(\alpha, \theta) &= \frac{1}{2\pi} \int_0^{2\pi} \frac{\exp(-ikR(\alpha, \theta, t) - ist)}{R(\alpha, \theta, t)} dt \\
&\cong \frac{1}{N_1} \sum_{\nu=1}^{N_1} \frac{\exp(-ikR(\alpha, \theta, t_\nu) - ist_\nu)}{R(\alpha, \theta, t_\nu)},
\end{aligned}
\tag{4.70}
$$

where $t = \varphi - \beta$, $t_\nu = \dfrac{(\nu - 0.5) \cdot 2\pi}{N_1}$.

To increase the speed of calculations, we can calculate DDFT by using the algorithm of fast Fourier transform (FFT).

Vasiliev [12] developed a method for calculating the functions $G_s(\theta, \alpha)$, which arise when we use the standard method of current integral equations, and, therefore, have singularities. Obviously, the use of this technique also in the case of functions with quasisingularities allows one to attain a higher accuracy.

Now we present formulas for calculating the functions $G_s(\theta, \alpha)$ according to this technique. We note that it is expedient to use these formulas only near a singularity (quasisingularity); formula (4.70) can be used in the other cases.

We introduce the notation

$$G_s^{\mathrm{I}} = \frac{1}{2\pi} \int_0^{2\pi} \frac{1}{R} \frac{\mathrm{d}}{\mathrm{d}R} \left(\frac{\exp(-iR)}{R} \right) \exp(-ist)\mathrm{d}t,$$

$$G_s^{\mathrm{II}} = \frac{1}{2\pi} \int_0^{2\pi} \frac{1}{R} \frac{\mathrm{d}}{\mathrm{d}R} \left[\frac{1}{R} \frac{\mathrm{d}}{\mathrm{d}R} \left(\frac{\exp(-iR)}{R} \right) \right] \exp(-ist)\mathrm{d}t.$$

The imaginary parts of the functions G_s, G_s^{I}, G_s^{II} can as usual be calculated by the rectangular formula, and the real parts, by the formulas [12]:

$$\mathrm{Re}\,G_s = A_s \sum_{m=0}^{\infty} \frac{(-1)^m R_0^{2m-1}}{m!2^{2m}\Gamma(m-s+1/2)} F\left(s-m+1/2, s+1/2, 2s+1, -x_0^2\right),$$

(4.71)

$$\mathrm{Re}\,G_S^{\mathrm{I}} = A_s \sum_{m=0}^{\infty} \frac{(-1)^m R_0^{2m-3}}{m!2^{2m-1}\Gamma(m-s-1/2)} F\left(s-m+3/2, s+1/2, 2s+1, -x_0^2\right),$$

(4.72)

$$\mathrm{Re}\,G_s^{\mathrm{II}} = A_s \sum_{m=0}^{\infty} \frac{(-1)^m R_0^{2m-5}}{m!2^{2m-2}\Gamma(m-s-3/2)} F\left(s-m+5/2, s+1/2, 2s+1, -x_0^2\right),$$

(4.73)

where $A_s = \dfrac{(-1)^s \Gamma(s+1/2)x_0^{2s}}{(2s)!}$, $R_0 = \sqrt{r^2 + \rho^2 - 2r\rho\sin\theta\sin\alpha}$,

$$x_0 = \frac{2\sqrt{r\rho\sin\theta\sin\alpha}}{R_0},$$

$$F\left(a, a+p, c, -x_0^2\right) = F\left(a+p, a, c, -x_0^2\right) = \frac{\Gamma(c)}{\Gamma(a)\Gamma(a+p)}$$

$$\left[\sum_{n=0}^{p-1} \frac{\Gamma(p-n)\Gamma(a+n)}{n!\Gamma(c-a-n)}(-1)^n x_0^{-2(n+a)} + \frac{1}{\Gamma(c-a)\Gamma(1-c+a)} \right.$$

$$\sum_{n=0}^{\infty} \frac{\Gamma(a+p+n)\Gamma(1-c+a+p+n)}{n!(p+n)!}(-1)^n x_0^{-2(n+p+a)} \cdot (2\ln x_0$$

$$\left. + \psi(1+p+n) + \psi(1+n) - \psi(a+p+n) - \psi(c-a-p-n))\right], \quad p \geq 0.$$

(4.74)

We note that in formulas (4.71)–(4.73), function F can only be calculated using formula (4.74) for $n = 0, 1$; for the other values of n, function F can be calculated by the recursive formula

$$(c-a)F(a-1,b,c,-x_0^2) + (2a-c+(a-b)x_0^2)F(a,b,c,-x_0^2) -$$
$$a(x_0^2+1)F(a+1,b,c,-x_0^2) = 0.$$

Despite the fact that it is necessary to use formulas (4.71)–(4.73) only near the singularities, i.e., for $x_0 > 5$ and $R_0 < 1$, they are too cumbersome and their use leads to an unnecessary increase in the computational time. Therefore, they cannot be optimal in this algorithm, and it is still more profitable to use the rectangular formula and to increase the accuracy of computations by some other methods that will be discussed below.

4.3.8 Study of Problems of Wave Diffraction at Compact Bodies

As an example of solving the problems of diffraction at compact scatterers by CBCM, we consider the problem of diffraction of a plane wave incident at the angles $\theta_0 = 0$ and $\varphi_0 = 0$ at the bodies whose geometry is characterized by the same ratio of the maximal to minimal size (1:10), i.e., the strongly prolate spheroid and cylinder with semiaxes $ka = 1$ and $kc = 10$, a cone of height $kc = 19$ with spherical base of radius $ka = 1$, and a trefoil of revolution determined by the equation $\rho(\theta) = c + a\cos 3\theta$ with the parameters $ka = 4.5$ and $kc = 5.5$, and with different impedances.

We consider these problems in the case where a boundary condition of the first type is satisfied on the scatterer, i.e., $Z = 0$. Figure 4.17a and b shows the respective scattering patterns and discrepancies for the spheroid (1), cylinder (2), and cone (3) on which the boundary condition of the first type is satisfied for $N = N_1 = 300$, $Q = 0$, and $k\delta = 10^{-3}$.

Now we consider the same problems but with the boundary condition of the third type for $Z = i\zeta$ and $Z = \zeta$. Figure 4.18a and b illustrates the respective scattering patterns and discrepancies for the spheroid (1), cylinder (2), and cone (3) in the case $Z = i\zeta$ for $N = N_1 = 400$, $Q = 0$, and $k\delta = 10^{-6}$. Figure 4.19a and b shows similar results for $Z = \zeta$.

Figure 4.20a and b shows the respective scattering patterns and discrepancies for the trefoil of revolution for different values of Z: 1—$Z = 0$, 2— $Z = 1000i\zeta$, and 3—$Z = \zeta$.

The optical theorem (3.24) was used to perform an additional verification of the pattern correctness in the case of ideally conducting scatterers. As an indicator of the optical theorem accuracy, we calculate the value of Δ_1 (3.25).

To verify the pattern correctness in the case of absorbing scatterers (with the impedance $Z = \zeta$), we verify the Ufimtsev theorem, which states that the integral cross-section of the absorbing scatterer is twice less than the integral cross-section of ideally conducting scatterers with the same shadow

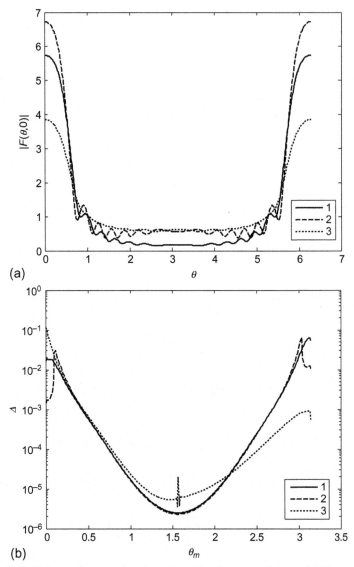

Figure 4.17 (a) Scattering patterns for spheroid, cylinder, and cone. (b) Discrepancies for spheroid, cylinder, and cone.

contour [92]. As an indicator of the Ufimtsev theorem accuracy we calculate the value of

$$\Delta_2 = \left| 2 - \frac{\sigma_S^0}{\sigma_S^i} \right|, \tag{4.75}$$

where σ_S^0 and σ_S^i are integral cross-sections of the ideally conducting and absorbing scatterers, respectively.

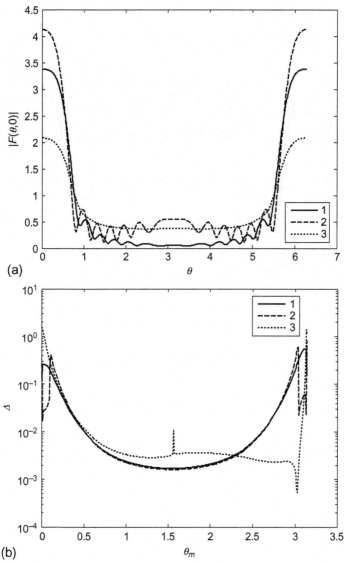

Figure 4.18 (a) Scattering patterns for the spheroid, cylinder, and cone, $Z=i\zeta$. (b) Discrepancies for spheroid, cylinder, and cone, $Z=i\zeta$.

Table 4.1 illustrates the optical theorem accuracy (Δ_1) for $Z=0$.

Table 4.2 illustrates the Ufimtsev theorem accuracy (Δ_2) for $Z=\zeta$.

One can see that the Ufimtsev theorem is satisfied with an appropriate accuracy even for such prolate bodies (except the cone) whose shadow contours surround a comparatively small area.

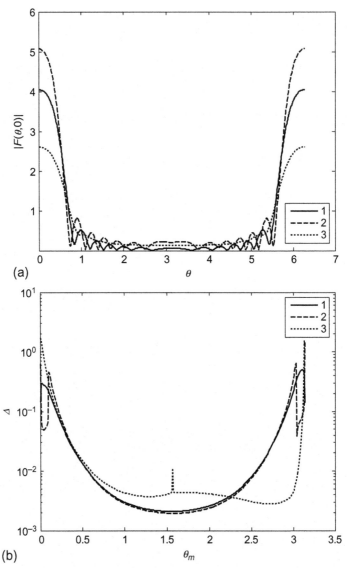

Figure 4.19 (a) Scattering patterns for spheroid, cylinder, and cone, $Z=\zeta$. (b) Discrepancies for spheroid, cylinder, and cone, $Z=\zeta$.

Now we consider the problems of diffraction at scatterers with variable impedance. We consider the problems of diffraction of a plane wave incident at the angles $\theta_0 = 0$ and $\varphi_0 = 0$ at circular cylinders of different sizes. One of the bases of each cylinder (on the side of the wave incidence) is coated by an absorbing material with impedance $Z = \zeta$, and the other parts

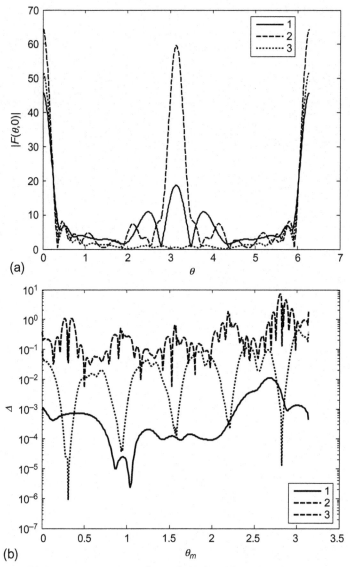

Figure 4.20 (a) Scattering patterns for trefoil of revolution, $Z=0$, $Z=1000i\zeta$, $Z=\zeta$. (b) Discrepancies for trefoil of revolution, $Z=0$, $Z=1000i\zeta$, $Z=\zeta$.

of the surface are ideally conducting (i.e., $Z=0$). We compare the scattering patterns for the cylinders with such a coating with the patterns of ideally conducting cylinders and absorbing cylinders of appropriate sizes with the impedance $Z=\zeta$.

Table 4.1 Optical theorem accuracy.

	Δ_1
Spheroid	0.0005897940
Cylinder	0.0007891644
Cone	0.0003681472
Trefoil	0.0002007159

Table 4.2 Ufimtsev theorem accuracy.

	Δ_2
Spheroid	0.2797938394
Cylinder	0.0925158857
Cone	0.8217138631
Trefoil	0.2008169751

Figure 4.21 shows the scattering patterns for the cylinders with semiaxes $ka = 4, kc = 8$ (a), $ka = 8, kc = 8$ (b), $ka = 8, kc = 4$ (c), $ka = 15, kc = 4$ (d), $ka = 15, kc = 2$ (e), $ka = 15, kc = 0.5$ (f), and $ka = 15, kc = 0.1$ (g). In these figures, we use the following notation: $1—Z = \zeta$, $2—Z = 0$, $\theta \leq \pi - \operatorname{arctg}(a/c)$, and $Z = \zeta$ otherwise, $3—Z = 0$, $4—Z = 0$, $\theta \leq \operatorname{arctg}(a/c)$ and $Z = \zeta$ otherwise. The results are obtained for $N = N_1 = 400$ and $Q = 0$.

Table 4.3 shows the results of the Ufimtsev theorem verification (Δ_2) in the case where the entire surface of the cylinders has the impedance $Z = \zeta$ and in the case where only the illuminated base of the cylinder has the impedance $Z = \zeta$ and the other surfaces are ideally conducting.

These results allow us to conclude that the impedance approximation ceases to be adequate for the disk thickness (cylinder height) $2c$ of the order of $0, 2\lambda$ and less. At the same time, in the case of thicker disks, the application of the absorbing coating to only one illuminated face allows us significantly to decrease the value of the field reflected from the source.

As examples of solving the vector problems of diffraction at compact scatterers by CBCM, we consider the problems of diffraction of a plane wave incident at the angles $\theta_0 = \frac{\pi}{2}$ and $\varphi_0 = 0$ at the bodies of the following geometry with the same ratio of the maximal to the minimal size (1:3): the ideally conducting spheroid and the cylinder with semiaxes $ka = 1$ and $kc = 3$ and the cone of height $kc = 5$ with spherical base of radius $ka = 1$.

Figure 4.22a–c shows the pattern components for spheroid (a), cylinder (b), and cone (c).

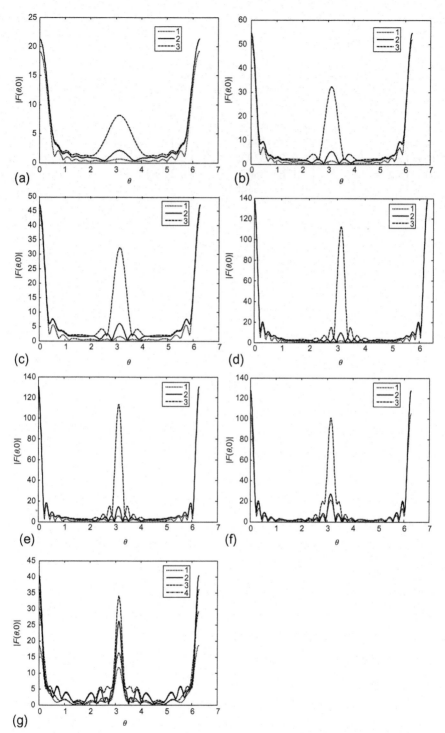

Figure 4.21 (a) Scattering patterns for cylinder with semiaxes $ka=4$, $kc=8$. (b) Scattering patterns for cylinder with semiaxes $ka=8$, $kc=8$. (c) Scattering patterns for cylinder with semiaxes $ka=8$, $kc=4$. (d) Scattering patterns for cylinder with semiaxes $ka=15$, $kc=4$. (e) Scattering patterns for cylinder with semiaxes $ka=15$, $kc=2$. (f) Scattering patterns for cylinder with semiaxes $ka=15$, $kc=0.5$. (g) Scattering patterns for cylinder with semiaxes $ka=15$, $kc=0.1$.

Table 4.3 Results of the Ufimtsev theorem verification.

Cylinder parameters	Absorbing cylinder	Cylinder with variable impedance
$ka = 4, kc = 8$	0.0172192692	0.6449646675
$ka = 8, kc = 8$	0.0086953753	0.5107656364
$ka = 8, kc = 4$	0.1133472859	0.4429342804
$ka = 15, kc = 4$	0.0490033932	0.2773196395
$ka = 15, kc = 2$	0.1767688676	0.2468326695
$ka = 15, kc = 0.5$	0.4288565480	0.3345994026
		0.0680601819[a]
$ka = 15, kc = 0.1$	2.3248395024	0.8272196614
		0.1308640661[a]

[a] The results obtained in the case where the absorbing material coats not only one of the cylinder bases but also its lateral surface.

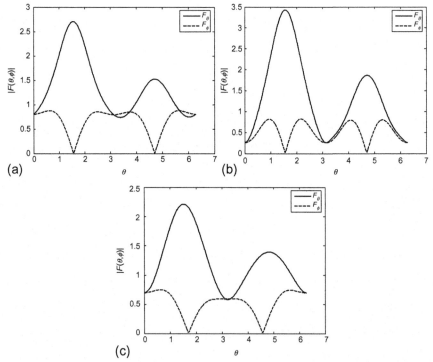

Figure 4.22 (a) Pattern components for spheroid. (b) Pattern components for cylinder. (c) Pattern components for cone.

4.3.9 Study of Problems of Wave Diffraction at Thin Screens

Now we consider the results of modeling of characteristics of wave radiation and scattering by antennas shaped as thin screens. As the initial field we consider not only the plane wave field (3.23) but also the field of a point source located at the scatterer axis of rotation:

$$U^0(\theta) = \frac{\exp\left(-ik\sqrt{r^2 + r_0^2 - 2rr_0\cos\theta}\right)}{k\sqrt{r^2 + r_0^2 - 2rr_0\cos\theta}}, \tag{4.76}$$

where r, θ are the spherical coordinates of the observation point and r_0 is the source coordinate.

We note that according to the existence theorem (Section 4.1.3), the surface S_δ must surround the screen surface S, but because of the local symmetry of the scattered field, an approximate boundary condition can be posed on the surface S_δ, which is also a screen moved away from S at a distance δ. The calculations show that for small $k\delta$, the result is independent of the side of the screen S on which the surface S_δ is chosen.

We calculate the scattering and radiation characteristics for a spherical mirror with $ka = 120$ and a parabolic mirror with focus at the origin and focal distance $kp = 120$ (the aperture diameter is equal to 120 in both cases), which are described by the equation

$$r(\theta) = \frac{p}{\cos^2 \theta/2},$$

in the following two cases: for the plane wave (3.23) is incident on the mirrors and for the field of a point source (4.76) on the axis of rotation with the coordinate $r_0 = a/2$ for the spherical mirror and with the coordinate $r_0 = 0$ for the parabolic mirror; $0 \le \theta \le \pi/6$ for both mirrors. Figure 4.23a and b shows the scattering patterns in the plane $\varphi = [0, \pi]$ and the boundary condition discrepancy for the plane wave (1 is the spherical mirror, 2 is the parabolic mirror) obtained for $Q = 0, N = 256$. Figure 4.24a and b shows the scattering patterns and the discrepancies for the point source, the notation is the same, and $Q = 0, N = 256$.

The main lobe of the scattering pattern in Fig. 4.23a is the shadow lobe whose width is approximately equal to the standard estimate: $\Delta\theta \approx \lambda/D$ [31], where D is the size of the mirror plane aperture.

Table 4.4 shows the results of the optical theorem verification in the case of a plane wave incident on the mirrors under study.

The directional patterns of a two-mirror Cassegrain-type antenna [93] were also calculated. The considered antenna consists of the main parabolic

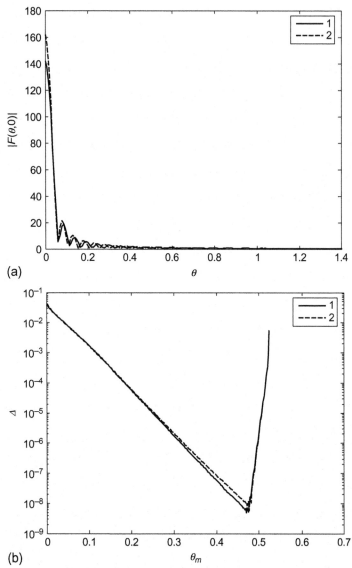

Figure 4.23 (a) Scattering patterns in the plane $\varphi = [0,\pi]$ for plane wave. (b) Boundary condition discrepancy for plane wave.

and auxiliary hyperbolic mirrors with common focus at the origin and with $0 \leq \theta \leq \pi/6$ for both mirrors. The hyperbolic mirror is described by the equation

$$\rho(\theta) = \frac{p-f}{(p/f)\cos^2(\theta/2) - 1}.$$

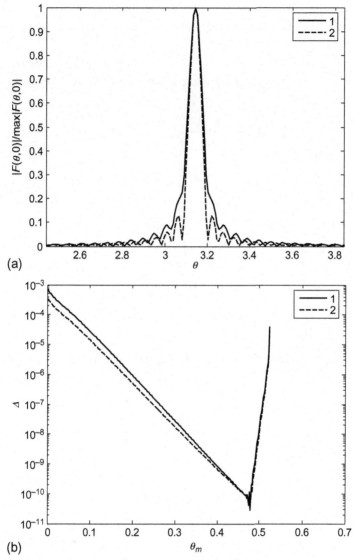

Figure 4.24 (a) Scattering patterns for the point source. (b) Discrepancies for the point source.

Table 4.4 Results of the optical theorem verification.

	Δ_1
Spherical mirror	0.0029273763
Parabolic mirror	0.0039126880

The initial field on the hyperboloid is assumed to be equal to the field of a point source located at the paraboloid vertex and equal to zero on the paraboloid.

Figure 4.25a and b shows the corresponding directional patterns (1—with the auxiliary mirror shadowing taken into account, 2—without the

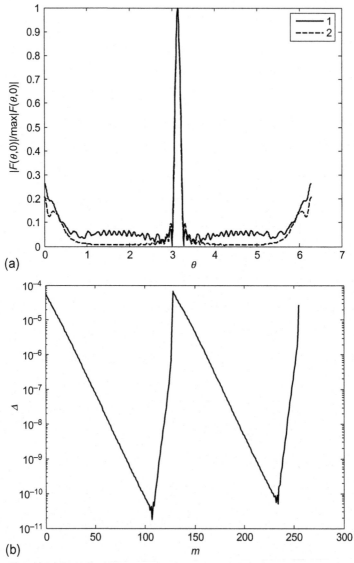

(a)

(b)

Figure 4.25 (a) Directional patterns for two-mirror Cassegrain-type antenna. (b) Solution discrepancy for two-mirror antenna.

shadowing taken into account) in the plane $\varphi = [0, \pi]$ and the discrepancy of the solution obtained for $kp = 60$, $kf = 6$, $Q = 0$, $N = 128$. One can see that the accuracy of the solution of the boundary-value problem is sufficiently high.

Now we consider the vector case. First, we consider the problem of excitation of a parabolic mirror with focal distance $kp = 30$ and opening angle $\pi/3$ by a point source (dipole) located on the paraboloid axis of rotation at the point $r_0 = 0$, $\theta_0 = \varphi_0 = 0$ and the problem of excitation of a spherical mirror of radius $ka = 30$ with the opening angle $\pi/3$. Thus, the initial field is given by the expression

$$\vec{E}^0 = \vec{p}\, U^0(r, \theta, \varphi), \tag{4.77}$$

where $\vec{p} = \vec{i}_x$ and the function U^0 is defined by formula (4.76).

Figure 4.26a and b, respectively, shows the normalized directional patterns in the plane $\varphi = [0, \pi]$ for F_θ^E and in the plane $\varphi = [\pi/2, 3\pi/2]$ for F_φ^E and the boundary condition discrepancy calculated at points between the collocation points, which were obtained for $N = N_1 = 128$ and $Q = 1$ (by using the algorithm without transition to discrete sources). Figure 4.27a and b shows similar results for the spherical mirror.

We note that, in the last case, the number N of functions used in the expansion of the required current is of the order of 10-15 L/λ, where L is the length of the mirror generatrix and λ is the wavelength. In this case, we obtain a quite admissible value of the boundary condition discrepancy (on a greater part of the mirror, its values do not exceed 10^{-2}-10^{-3}). Approximately the same accuracy with the same recommendations about the number of basic functions is obtained in the classical method of current integral equations [8,11] by using a well-developed but still much more complicated technique.

Now we consider the problem of plane wave diffraction at the above-described thin screens. Figure 4.28a and b shows the respective normalized scattering patterns in the plane $\varphi = [0, \pi]$ for F_θ^E and in the plane $\varphi = [\pi/2, 3\pi/2]$ for F_φ^E and the boundary condition discrepancy for the parabolic mirror obtained for $N = 300$, $N_1 = 64$, and $Q = 1$ (by using the algorithm with transition to discrete sources). Figure 4.29a and b shows similar results for the spherical mirror.

4.3.10 Asymptotic Solution of Diffraction Problems by the Method of Continued Boundary Conditions

It is very difficult to solve the problem of diffraction at bodies whose dimensions are much greater than the wavelength. The approaches used in this case

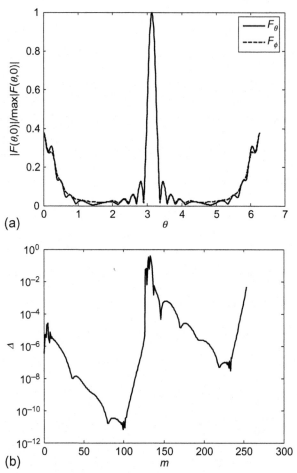

Figure 4.26 (a) Normalized directional patterns for parabolic mirror. (b) Boundary condition discrepancy for parabolic mirror.

can conditionally be divided into two groups, asymptotic [38] and numerical [11]. When the problem of diffraction at scatterers whose dimensions are much greater than the wavelength is solved numerically, one meets quite obvious computational difficulties related, in particular, to the very large sizes of systems of algebraic equations to which the initial boundary-value problem is reduced. The so-called hybrid approach [3,134] seems to be sufficiently obvious. In this approach, in the integral equation for the current on the scatterer surface, the required function (current) is represented as the sum of two terms, a known quantity (geometric-optical current) and an unknown part localized near the "light-shadow" interface. In Ref. [134],

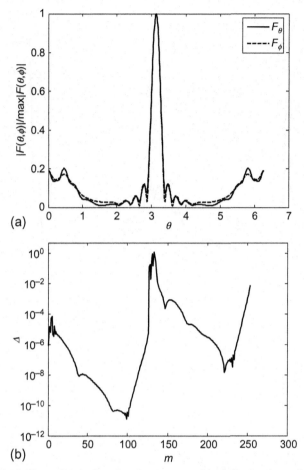

Figure 4.27 (a) Normalized directional patterns of spherical mirror. (b) Boundary condition discrepancy for spherical mirror.

the method of continued boundary conditions was used to determine the unknown addition to the geometric-optical current in the two-dimensional diffraction problem. It is of interest to generalize the approach proposed in Ref. [134] to the case of three-dimensional diffraction problems.

We consider the three-dimensional scalar problem of diffraction at an axisymmetric scatterer with the boundary S. For the definiteness, we assume that the Dirichlet condition is satisfied on the boundary S, i.e., $\beta = 0$ in relation (4.2). The CBCM integral equation in this case has the form (see (4.6))

$$U^0\left(\vec{r}\right) - \int_S \frac{\partial U\left(\vec{r}'\right)}{\partial n'} G_0\left(\vec{r},\vec{r}'\right) ds' = 0, \quad M\left(\vec{r}\right) \in S_\delta. \tag{4.78}$$

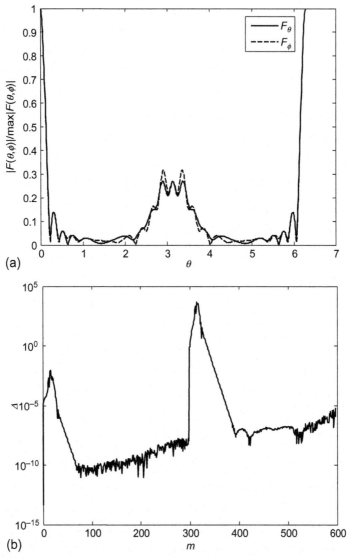

Figure 4.28 (a) Normalized scattering patterns for parabolic mirror. (b) Boundary condition discrepancy for parabolic mirror.

Passing to spherical coordinates and introducing the notation

$$\left[\frac{\partial U}{\partial r'}\rho(\theta') - \frac{\partial U}{\partial \theta'}\frac{\rho'(\theta')}{\rho(\theta')}\right]\Bigg|_{r'=\rho(\theta')} \equiv I(\theta', \varphi'),$$

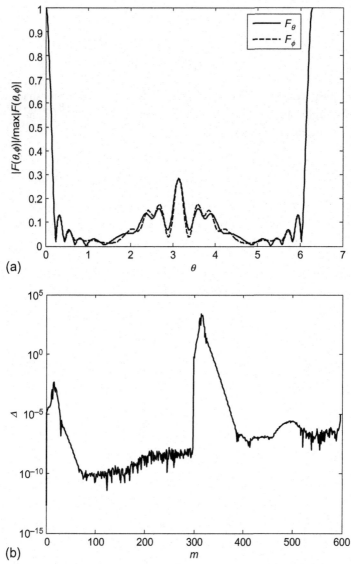

Figure 4.29 (a) Normalized scattering patterns for spherical mirror. (b) Boundary condition discrepancy for spherical mirror.

where $r = \rho(\theta)$ is the equation of the surface S, we rewrite Equation (4.78) as

$$U^0(\vec{r}) - \int_0^{2\pi} \int_0^{\pi} I(\theta', \varphi') G_0(\vec{r}, \vec{r}') \sin\theta' \, d\theta' \, d\varphi' = 0, \quad M(\vec{r}) \in S_\delta. \quad (4.79)$$

In this equation, we further assume [134] that

$$I(\theta', \varphi') = I^0(\theta', \varphi') \equiv 2\left[\frac{\partial U^0}{\partial r'}\rho(\theta') - \frac{\partial U^0}{\partial \theta'}\frac{\rho'(\theta')}{\rho(\theta')}\right]\Bigg|_{r'=\rho(\theta')} \tag{4.80}$$

on the illuminated part of the scatterer and $I(\theta', \varphi') = 0$ in the shadow; and near the "light-shadow" interface, we set $I(\theta', \varphi') = I^1(\theta', \varphi')$, where $I^1(\theta', \varphi')$ is the unknown function.

Let

$$U^0(\vec{r}) = \exp\{-ikr[\sin\theta\sin\theta_0\cos(\varphi - \varphi_0) + \cos\theta\cos\theta_0]\}$$

be a plane wave propagating at the angles θ_0, φ_0. For simplicity, we consider the case where $\theta_0 = 0$. Then we have $I(\theta', \varphi') = I(\theta')$ due to the axial symmetry. Now according to the above concept, we have

$$I(\theta') = \begin{cases} I^0(\theta'), & \pi - \alpha \le \theta' \le \pi, \\ I^1(\theta'), & \alpha \le \theta' \le \pi - \alpha, \\ 0, & 0 \le \theta' < \alpha. \end{cases} \tag{4.81}$$

As a result, equation (4.79) becomes

$$\begin{aligned} U^0(\vec{r}) &- \int_\alpha^{\pi-\alpha} I^1(\theta') \int_0^{2\pi} G_0(\vec{r}, \vec{r}')\sin\theta' d\theta' d\varphi' \\ &- \int_{\pi-\alpha}^\pi I^0(\theta') \int_0^{2\pi} G_0(\vec{r}, \vec{r}')\sin\theta' d\theta' d\varphi' = 0, \quad M(\vec{r}) \in S_\delta. \end{aligned} \tag{4.82}$$

According to (4.80), in this equation we have

$$I^0(\theta') = -2ik\rho(\theta')\left[\cos\theta' + \frac{\rho'(\theta')}{\rho(\theta')}\sin\theta'\right]e^{-ik\rho(\theta')\cos\theta'}. \tag{4.83}$$

The parameter α is chosen in numerical experiments where, for example, the optical theorem (3.24) is chosen as the correctness criterion of the obtained results.

Determining $I^1(\theta')$ from Equation (4.82), we calculate the scattering pattern by the formula

$$\begin{aligned} f(\theta, \varphi) &= -\frac{1}{2}\int_\alpha^{\pi-\alpha} I^1(\theta')J_0(k\rho(\theta')\sin\theta\sin\theta')e^{ik\rho(\theta')\cos\theta\cos\theta'}\sin\theta' d\theta' \\ &- \frac{1}{2}\int_{\pi-\alpha}^\pi I^0(\theta')J_0(k\rho(\theta')\sin\theta\sin\theta')e^{ik\rho(\theta')\cos\theta\cos\theta'}\sin\theta' d\theta'. \end{aligned}$$

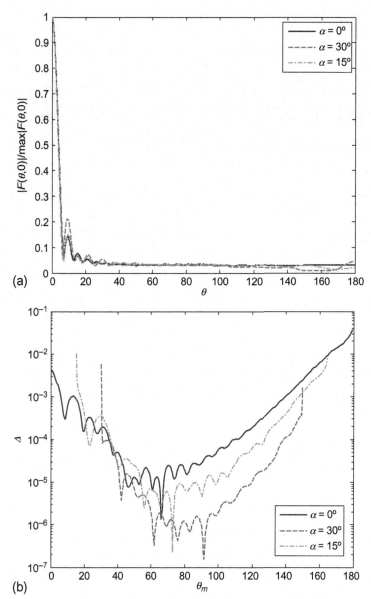

Figure 4.30 (a) Scattering pattern for a sphere of radius $ka = 30$. (b) Boundary condition discrepancy for a sphere of radius $ka = 30$.

Now let us discuss the results of numerical studies. Figure 4.30a illustrates the results of the scattering pattern calculation for a sphere of radius $ka = 30$ and for three values of the angle α: $0°$, $15°$, and $30°$. The pattern for $\alpha = 0$ is assumed to be standard. Integral equation (4.82) is solved by the methods of discrete sources and collocation. As a result, the problem is reduced to solving an $N \times N$ system of algebraic equations. The results shown in Fig. 4.30a were obtained for $N = 400, k\delta = 0,001$. The optical theorem accuracy for $\alpha = 0$ is 0.001766332. One can see that the pattern is close to the standard one for $\alpha = 15°$. For $\alpha = 30°$, the difference is more noticeable, but there is practically no difference in the region of the pattern main lobe.

Figure 4.30b shows the graphs of the boundary condition discrepancy on the surface S_δ calculated at the middle points between the collocation points. One can see that the discrepancy is quite acceptable for all values of the parameter α. Moreover, the discrepancy for $\alpha = 15°$ is noticeably lower than that for $\alpha = 0$, and the decrease in the discrepancy for $\alpha = 30°$ is much greater. This testifies that to obtain the same accuracy, the number of discrete sources must be decreased as the parameter α increases.

Figure 4.31a and b shows the results of similar calculations for a sphere of radius $ka = 70$ for $N = 600$. One can see that, in this case, the parameter α can already take much greater values for an acceptable accuracy of calculations.

Figure 4.32a and b shows the respective results of calculations of the scattering pattern and the discrepancy in the problem of the plane wave diffraction at a prolate spheroid with parameters $ka = 30, kc = 50$, where c is the semiaxis directed along the axis Oz. The calculations were performed for $N = 500$.

Finally, Fig. 4.33a and b shows the results of similar calculations for an oblate ($ka = 50, kc = 30$) spheroid. In this case, the number of discrete sources in the calculations varies with the parameter α: for $\alpha = 0$, this number is $N = 600$, for $\alpha = 30°$, $N = 500$, and finally for $\alpha = 45°$, $N = 400$. One can see that despite a noticeable decrease in the size of the algebraic system, the accuracy of the pattern calculations is significantly greater than in the case of a prolate spheroid, which agrees well with the concepts of geometrical optics. Moreover, as follows from Fig. 4.33b, the number of discrete sources in the "light-shadow" region for $\alpha = 30°$ and $\alpha = 45°$ can be further decreased without loss of accuracy of calculations.

Summarizing the above considerations, we conclude that the proposed scheme for obtaining asymptotic solutions of diffraction problems at scatterers of large dimensions is quite efficient, despite the fact that it is based on an approximate approach, namely, the method of continued boundary conditions. This approach can also be generalized to solving the vector problems of wave scattering.

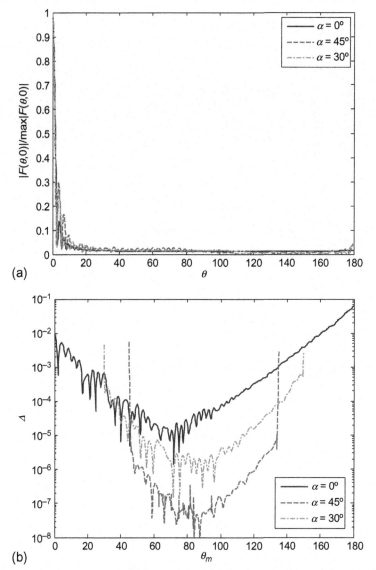

Figure 4.31 (a) Scattering pattern for a sphere of radius $ka = 70$. (b) Boundary condition discrepancy for a sphere of radius $ka = 70$.

4.4 MODIFIED METHOD OF CONTINUED BOUNDARY CONDITIONS

To conclude this chapter, we note that the CBCM formulation can principally be improved by using the expansion of the diffraction field in the Taylor series in a neighborhood of the boundary and preserving, for example, the

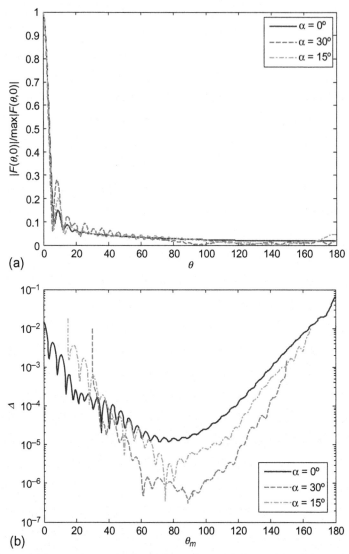

Figure 4.32 (a) Scattering pattern for prolate spheroid with parameters $ka = 30$, $kc = 50$. (b) Boundary condition discrepancy for prolate spheroid with parameters $ka = 30$, $kc = 50$.

first two terms of the expansion. Such an idea was realized by Bogomolov and Kleev in Ref. [135] with the aim of increasing the CBCM accuracy. This approach was called the modified method of continued boundary conditions (MCBCM) in Ref. [135]. But more precise studies [136] inspired by Bogomolov and Kleev [135] showed that the desired increase in the

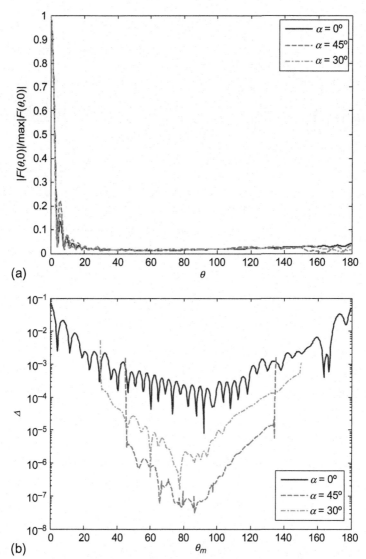

Figure 4.33 (a) Scattering pattern for oblate ($ka = 50$, $kc = 30$) spheroid. (b) Boundary condition discrepancy for oblate ($ka = 50$, $kc = 30$) spheroid.

accuracy can also be obtained in the framework of CBCM by decreasing the parameter δ.

We consider the diffraction problem (4.1) and (4.2) and replace the boundary condition (4.2) for $\alpha = 1$, $\beta = 0$ by an approximate condition with the first-order terms in the Taylor expansion of $U(\vec{r})$ taken into account:

$$U^0(\vec{r}) + U^1(\vec{r})\big|_{S_\delta} = \delta \frac{\partial}{\partial n}\left(U^0(\vec{r}) + U^1(\vec{r})\right)\bigg|_{S_\delta} \qquad (4.84)$$

Substituting representation (4.4) into the boundary condition (4.84), we obtain the Fredholm integral equation of the first type with smooth kernel

$$\int_S I(\vec{r}') \cdot \left(\frac{\partial}{\partial n} G_0(\vec{r},\vec{r}') - G_0(\vec{r},\vec{r}')\right) ds' = U^0(\vec{r}) - \delta \frac{\partial U^0(\vec{r})}{\partial n}, \qquad (4.85)$$

where $M(\vec{r}) \in S_\delta, N(\vec{r}') \in S$, and $G_0(\vec{r},\vec{r}') = \frac{1}{4i}H_0^{(2)}(k|\vec{r} - \vec{r}'|)$.

As previously noted, the use of MCBCM allows one to decrease the intrinsic error of the solution. To determine to what extent the error can be decreased compared with the standard CBCM, we consider the problem of diffraction at a circle of radius a. In this case, the diffraction problem can be solved analytically. As S_δ we take a circle of radius $a + \delta$.

We expand the incident and scattered fields in the Fourier series

$$U^0(\varphi) = \sum_{n=-\infty}^{\infty} (-i)^n J_n(kr)e^{in\varphi}, \qquad (4.86)$$

$$U^1(\varphi) = \sum_{n=-\infty}^{\infty} c_n H_n^{(2)}(kr)e^{in\varphi}. \qquad (4.87)$$

We substitute these expansions into the exact boundary condition (4.2) for $\alpha = 1$, $\beta = 0$,

$$\sum_{i=1}^{n} \left[(-i)^n J_n(ka) + c_n H_n^{(2)}(ka)\right]e^{in\varphi} = 0, \qquad (4.88)$$

and obtain

$$c_n = -(-i)^n \frac{J_n(ka)}{H_n^{(2)}(ka)}. \qquad (4.89)$$

Further, we substitute these expansions in the CBCM boundary condition (4.3) for $\alpha = 1$, $\beta = 0$,

$$\sum_{i=1}^{n} \left[(-i)^n J_n(k(a+\delta)) + c_n^{CBCM} H_n^{(2)}(k(a+\delta))\right]e^{in\varphi} = 0, \qquad (4.90)$$

and obtain

$$c_n^{CBCM} = -(-i)^n \frac{J_n(k(a+\delta))}{H_n^{(2)}(k(a+\delta))}. \qquad (4.91)$$

Finally, we substitute these expansions in the MCBCM boundary condition (4.84):

$$\sum_{n=-\infty}^{\infty} \left[(-i)^n J_n(k(a+\delta)) + c_n^{\text{MCBCM}} H_n^{(2)}(k(a+\delta)) \right] e^{in\varphi}$$
$$= k\delta \sum_{n=-\infty}^{\infty} \left[(-i)^n J_n'(k(a+\delta)) + c_n^{\text{MCBCM}} H_n^{(2)\prime}(k(a+\delta)) \right] e^{in\varphi} \qquad ; \qquad (4.92)$$

we obtain

$$c_n^{\text{MCBCM}} = -(-i)^n \frac{J_n(k(a+\delta)) - k\delta J_n'(k(a+\delta))}{H_n^{(2)}(k(a+\delta)) - k\delta H_u^{(2)\prime}(k(a+\delta))}. \qquad (4.93)$$

Now we estimate the difference between the expansion coefficients of the unknown function (scattered field) obtained by CBCM and MCBCM and the exact values of these coefficients. For this, we estimate the differences $c_n - c_n^{\text{CBCM}}$ and $c_n - c_n^{\text{MCBCM}}$:

$$c_n - c_n^{\text{CBCM}} = -(-i)^n \frac{J_n(k(a+\delta)) H_n^{(2)}(ka) - J_n(ka) H_n^{(2)}(k(a+\delta))}{H_n^{(2)}(ka) H_n^{(2)}(k(a+\delta))}.$$

With the relations $J_n(k(a+\delta)) \simeq J_n(ka) + k\delta J_n'(ka)$, $H_n^{(2)}(k(a+\delta)) \simeq H_n^{(2)}(ka) + k\delta H_n^{(2)}\prime(ka)$ taken into account, we obtain

$$c_n - c_n^{\text{CBCM}} = -(-i)^n \frac{k\delta \left[J_n'(ka) H_n^{(2)}(ka) - J_n(ka) H_n^{(2)\prime}(ka) \right]}{H_n^{(2)}(ka) \left[H_n^{(2)}(ka) + k\delta H_n^{(2)\prime}(ka) \right]} + O\big((k\delta)^2\big)$$
$$\sim k\delta.$$

$$(4.94)$$

Similarly, we can show that

$$c_n - c_n^{\text{MCBCM}} \sim (k\delta)^2. \qquad (4.95)$$

Thus, the use of MCBCM theoretically allows one to decrease the intrinsic error of the solution by an order of magnitude.

To compare the accuracy of CBCM and MCBCM numerically, we again consider the problem of diffraction at a circle, because the results obtained in this case can be compared with the analytic solution. We determine the error of the pattern calculations by the formulas

$$\Delta_g^{\text{MCBCM}} = \max_{\varphi \in [0, 2\pi]} \left| g(\varphi) - g_{\text{MCBCM}}(\varphi) \right|, \qquad (4.96)$$

$$\Delta_g^{CBCM} = \max_{\varphi \in [0, 2\pi]} \left| g(\varphi) - g_{CBCM}(\varphi) \right|, \tag{4.97}$$

where $g_{CBCM}(\varphi) = \sum_{n=1}^{N} I_n^{CBCM} e^{ik\rho(\varphi_n)\cos(\varphi - \varphi_n)}$, $g_{MCBCM}(\varphi) = \sum_{n=1}^{N} I_n^{MCBCM} e^{ik\rho(\varphi_n)\cos(\varphi - \varphi_n)}$, I_n^{CBCM} and I_n^{MCBCM} are the currents obtained by numerically solving the CBCM integral equations and the MCBCM integral equations, respectively, and $g(\varphi)$ is the pattern for the exact solution

$$g(\varphi) = - \sum_{n=-M}^{M} (-i)^n \frac{J_n(ka)}{H_n^{(2)}(ka)} e^{ikacos(\varphi - \varphi_n)}. \tag{4.98}$$

We consider the problem of diffraction of the plane wave (3.44) incident at the angle $\varphi_0 = 90°$ at a circle of radius $ka = 2$. Table 4.5 shows the values of the error in the CBCM and MCBCM calculations of the pattern for different N and $k\delta$.

One can see that if the discretization step $\frac{2\pi}{N}$ is greater than $k\delta$, i.e., if the statement of the problem in the discrete version is equivalent to the exact statement, then MCBCM has a somewhat greater error than the errors obtained by CBCM, but the order of the errors is approximately the same. This is quite expectable, because in the framework of such a discretization, the CBCM and MCBCM integral equations practically coincide, but because of the greater complexity of the kernel in the MCBCM integral equation and a stronger quasisingularity in this kernel, the accuracy of calculations for a fixed N in MCBCM is somewhat lower. But if the discretization step is less than $k\delta$, then the MCBCM error is approximately by an

Table 4.5 Numerical comparison of the accuracy of the methods

N	$2\pi/N$	$k\delta$	Δ_g^{CBCM}	Δ_g^{MCBCM}
50	0.126	10^{-2}	0.01494	0.03025
200	0.031	10^{-2}	0.00146	0.00283
700	0.009	10^{-2}	0.00267	1.01376×10^{-4}
50	0.126	10^{-3}	0.05228	0.07215
200	0.031	10^{-3}	0.00625	0.00930
700	0.009	10^{-3}	6.72187×10^{-4}	0.00159
1000	0.006	10^{-3}	2.33777×10^{-4}	9.21070×10^{-4}
6500	0.001	10^{-3}	2.65987×10^{-4}	1.40802×10^{-5}
200	0.031	10^{-4}	0.01323	0.01640
700	0.009	10^{-4}	0.00260	0.00340
1000	0.006	10^{-4}	0.00161	0.00217
6500	0.001	10^{-4}	7.93407×10^{-5}	1.76488×10^{-4}

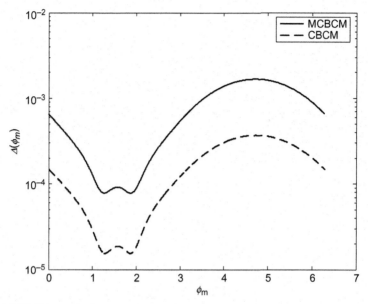

Figure 4.34 Discrepancies.

order of magnitude less than the CBCM error, as was predicated by the analytic comparison. But the calculation algorithm is not efficient for $\frac{2\pi}{N} < k\delta$, because in this case, even for $k\delta = 10^{-3}$, the value of N must exceed 6000 independently of the scatterer dimensions.

As an additional criterion for the solution accuracy, we calculate the boundary condition discrepancy at points between the collocation points. Figure 4.34 shows the discrepancies for the considered problem obtained by CBCM and MCBCM for $k\delta = 10^{-2}$, $N = 700$, and $ka = 2$.

The MCBCM discrepancy is worse than the CBCM discrepancy, which can be explained by the fact that the kernel of the MCBCM integral equation contains the derivative of the Hankel function.

CHAPTER 5

Pattern Equation Method

The pattern equation method (PEM), first proposed in Kyurkchan [51,94], was then generalized to a wide range of wave diffraction and propagation problems (see [42,44,63,64,66,95–131]). We briefly illustrate the main idea of the method with an example of two-dimensional problems of wave diffraction at a compact scatterer [51,94].

5.1 SOLUTION OF TWO-DIMENSIONAL PROBLEM OF DIFFRACTION AT A COMPACT SCATTERER USING THE PATTERN EQUATION METHOD

5.1.1 Integral-Operator Equation

We consider the case of E-polarization where the single component of the vector $\vec{E}$ is parallel to the axis z of a cylindrical body with a smooth directrix S. The case of H-polarization can be considered quite similarly. We introduce the notation $E_z \equiv U = U^0 + U^1$, where U^0 is the initial field and U^1 is the scattered field. We assume that the inclusion $\overline{B}_0 \subseteq \overline{D}$, where D is the domain bounded by S, holds for the convex envelope $\overline{B}_0$ of singularities of the diffraction field U^1. As shown in Chapter 1, the field $U^1(r, \varphi)$ can be represented by the integral (1.35) everywhere in $\mathbb{R}^2/\overline{B}_0$.

We assume that the impedance boundary condition $E_z|_S = Z H_\tau|_S$ is satisfied on S, i.e.,

$$\left(U - \frac{Z}{ik\zeta} \frac{\partial U}{\partial n} \right) \Bigg|_S = 0, \tag{5.1}$$

where $\zeta = \sqrt{\mu_0/\varepsilon_0}$. In this case, because (see (1.43))

$$f(\alpha) = \frac{i}{4} \int_S \left\{ \frac{\partial U(\vec{r})}{\partial n} - U(\vec{r}) \frac{\partial}{\partial n} \right\} \exp[ikr\cos(\alpha - \varphi)] ds,$$

we can use (5.1) and the relations

$$\frac{\partial}{\partial n} = \left(\rho^2(\varphi) + \rho'^2(\varphi)\right)^{-1/2} \left(\rho(\varphi)\frac{\partial}{\partial r} - \frac{\rho'(\varphi)}{\rho(\varphi)}\frac{\partial}{\partial \varphi}\right), \quad ds = \left(\rho^2 + \rho'^2\right)^{1/2} d\varphi,$$

where $r = \rho(\varphi)$ is the equation of the contour S, to obtain

$$f(\alpha) = \int_0^{2\pi} v(\varphi) \left\{ 1 - \frac{iW}{\sqrt{\rho^2 + \rho'^2}} [\rho(\varphi)\cos(\alpha - \varphi) - \rho'(\varphi)\sin(\alpha - \varphi)] \right\}$$
$$\times \exp[ik\,\rho(\varphi)\cos(\alpha - \varphi)]d\varphi$$

(5.2)

In (5.2), we introduced the notation

$$v(\varphi) = \frac{i}{4}\left(\rho(\varphi)\frac{\partial U}{\partial r} - \frac{\rho'(\varphi)}{\rho(\varphi)}\frac{\partial U}{\partial \varphi}\right)\bigg|_{r=\rho(\varphi)}, \quad Z = i\zeta W. \qquad (5.3)$$

We denote the contour of integration in (1.35) by Γ. Then we substitute representation (1.35) into (5.3) and, after elementary transformations [94], obtain

$$v(\varphi) = v^0(\varphi) + \frac{1}{\pi}\int_\Gamma \frac{k}{4}[\rho(\varphi)\cos\psi - \rho'(\varphi)\sin\psi]f(\varphi + \psi)\exp[-ik\,\rho(\varphi)\cos\psi]d\psi.$$

Here $v^0(\varphi)$ is an expression of the form (5.3), where $U(r,\varphi)$ must be replaced by the initial field $U^0(r,\varphi)$.

Now, using (5.2), we easily obtain the integral-operator equation of the second type for the scattering pattern

$$f(\alpha) = f^0(\alpha) + \frac{1}{\pi}\int_0^{2\pi}\int_\Gamma f(\varphi + \psi)\frac{k}{4}[\rho(\varphi)\cos\psi - \rho'(\varphi)\sin\psi]$$

$$\times \left\{ 1 - \frac{iW}{\sqrt{\rho^2 + \rho'^2}}[\rho(\varphi)\cos(\alpha - \varphi) - \rho'(\varphi)\sin(\alpha - \varphi)] \right\} \qquad (5.4)$$

$$\times \exp\{-ik\,\rho(\varphi[\cos\psi - \cos(\alpha - \varphi)])\}d\psi d\varphi$$

where

$$f^0(\alpha) = \int_0^{2\pi} v^0(\varphi) \left\{ 1 - \frac{iW}{\sqrt{\rho^2 + \rho'^2}} [\rho(\varphi)\cos(\alpha - \varphi) - \rho'(\varphi)\sin(\alpha - \varphi)] \right\}$$
$$\times \exp[ik\,\rho(\varphi)\cos(\alpha - \varphi)]d\varphi$$

(5.5)

is a known function.

Equation (5.4) is solvable (this follows from its derivation) in the class of 2π-periodic functions $f(\psi)$, which can analytically be continued to the entire complex plane $\psi = \psi' + i\psi''$ and have the asymptotics as $|\psi''| \to \infty$

$$f(\psi) = f^E(w_\pm)\left(1 + O\left(e^{-|\psi''|}\right)\right),$$

where $f^E(w_\pm)$ is an entire function of a finite degree of complex variables $w_\pm = e^{|\psi''|}e^{\pm i\psi'}$, and the inclusion $\bar{B} \subseteq \bar{B}_0$ holds for the convex envelope $\bar{B}$ of singularities of the field corresponding to the function $f(\psi)$.

The problem of uniqueness of the solution of Equation (5.4) will be considered later.

Now we rewrite Equation (5.4) in the special case of circular cylinder $\rho(\varphi) = a$ as

$$f(\alpha) = f^0(\alpha) + \frac{ka}{4\pi} \int_0^{2\pi} \int_\Gamma [1 - iW\cos(\alpha - \varphi)]f(\varphi + \psi)$$
$$\times \exp\{ika[\cos(\alpha - \varphi) - \cos\psi]\}\cos\psi\,d\psi\,d\varphi$$

(5.6)

and

$$f^0(\alpha) = \int_0^{2\pi} v^0(\varphi)[1 - iW\cos(\alpha - \varphi)]\exp[ika\cos(\alpha - \varphi)]d\varphi.$$

Let

$$U^0(\vec{r}) = \frac{1}{4i}H_0^{(2)}\left(k|\vec{r} - \vec{r}_0|\right),$$

then

$$f^0(\alpha) = \frac{\pi ka}{8} \sum_{n=-\infty}^{\infty} i^n J_n'(ka)H_n^{(2)}(kr_0)\left[J_n(ka) - WJ_n'(ka)\right]e^{in(\alpha - \varphi_0)}.$$

We iterate the kernel of Equation (5.6) and obtain the expression for the resolvent

$$R(\alpha; \varphi, \psi) = \frac{2i}{\pi ka} \sum_{n=-\infty}^{\infty} \frac{\left[J_n(ka) - WJ'_n(ka) \right] e^{in(\alpha - \varphi + \pi/2)}}{\left[H_n^{(2)}(ka) - WH_n^{(2)\prime}(ka) \right] J'_n(ka)} e^{-ika\cos\psi} \cos\psi.$$

(5.7)

The series of successive approximations converges for

$$\left| \frac{i\pi ka}{2} \left[J_n(ka) - WJ'_n(ka) \right] H_n^{(2)\prime}(ka) \right| < 1,$$

but expression (5.7) is meaningful for all ka except the values for which the denominators of the terms in (5.7) are zero. Thus, as follows from (5.7), the poles of the resolvent are the zeros of the expressions

$$H_n^{(2)}(ka) - WH_n^{(2)\prime}(ka).$$

Using the relation

$$f(\alpha) = f^0(\alpha) + \frac{ka}{4\pi} \int_0^{2\pi} \int_\Gamma R(\alpha; \varphi, \psi) f^0(\varphi + \psi) \, d\psi \, d\varphi,$$

we obtain the well-known expression

$$f(\alpha) = \frac{i}{4} \sum_{n=-\infty}^{\infty} i^n \frac{J_n(ka) - WJ'_n(ka)}{H_n^{(2)}(ka) - WH_n^{(2)\prime}(ka)} H_n^{(2)}(kr_0) e^{in(\alpha - \varphi_0)}.$$

5.1.2 Algebraization of the Problem

Evidently, Equation (5.4) cannot be solved easily in the general case, and, therefore, it is expedient to algebraize it. The most justified method is then to use the Fourier expansions of the functions $f(\alpha)$ and $f^0(\alpha)$

$$f(\alpha) = \sum_{n=-\infty}^{\infty} a_n e^{in\alpha}, \quad f^0(\alpha) = \sum_{n=-\infty}^{\infty} a_n^0 e^{in\alpha}.$$

(5.8)

Substituting (5.8) into (5.4), we obtain an algebraic system for the coefficients a_n of the form

$$a_n = a_n^0 + \sum_{m=-\infty}^{\infty} G_{nm} a_m,$$

(5.9)

where the coefficients a_n^0 are known (see (5.5)) and are determined by the form of the function $v^0(\varphi)$, i.e., by the initial field, and

$$G_{nm} = \frac{1}{4} \int_0^{2\pi} \left\{ J_n(k\rho) - \frac{iW}{\sqrt{\rho^2 + \rho'^2}} \left[i\rho(\varphi)J_n'(k\rho) + \frac{n\rho'(\varphi)}{k\rho(\varphi)} J_n(k\rho) \right] \right\}$$

$$\times \left[ik\,\rho(\varphi)H_m^{(2)\prime}(k\,\rho) + m\frac{\rho'(\varphi)}{\rho(\varphi)} H_m^{(2)}(k\rho) \right] \exp[i(m-n)(\varphi - \pi/2)]\,\mathrm{d}\varphi$$

$$(5.10)$$

Thus, in contrast to the algebraic systems traditionally used in problems of diffraction at cylindrical bodies (including the systems that arise when the integral equation is algebraized with respect to the current), the matrix elements of system (5.9) are expressed in terms of integrals of multiplicity one, and this is an undeniable advantage. Moreover, we note that the above-mentioned "current" equation cannot be algebraized so simply.

Let us estimate the asymptotics of the matrix elements G_{mn}. To simplify the transformations, we consider the case $W = 0$.

For $n \gg k\rho$ [35,40], we have

$$J_n(k\rho) \cong \frac{1}{n!}\left(\frac{k\rho}{2}\right)^n - \frac{1}{(n+1)!}\left(\frac{k\rho}{2}\right)^{n+2},$$

$$H_n^{(2)}(kr) \cong \frac{i}{\pi}\left[(n-1)!\left(\frac{2}{k\rho}\right)^n + (n-2)!\left(\frac{2}{k\rho}\right)^{n-2} \right],$$

$$(5.11)$$

and hence as $m \to \infty$, $n \to \infty$, we obtain

$$G_{mn} \cong \left(\tfrac{ik}{2}\right)^{m-n} \frac{n!}{4\pi m!} \int_0^{2\pi} \left(\rho(\varphi)e^{-i\varphi}\right)^{m-n} \left(1 + i\frac{\rho'(\varphi)}{\rho(\varphi)}\right) \mathrm{d}\varphi = \frac{1}{2}\delta_{mn},$$

$$G_{-m,-n} \cong \left(\tfrac{ik}{2}\right)^{m-n} \frac{n!}{4\pi m!} \int_0^{2\pi} \left(\rho(\varphi)e^{i\varphi}\right)^{m-n} \left(1 - i\frac{\rho'(\varphi)}{\rho(\varphi)}\right) \mathrm{d}\varphi = \frac{1}{2}\delta_{mn}.$$

Here we used the relations, which can easily be verified (for example, by applying the change $\zeta^{\pm} = \rho(\varphi)e^{\pm i\varphi}$ and integrating in the complex plane ζ),

$$\frac{1}{2\pi} \int_0^{2\pi} \left(\rho(\varphi)e^{\pm i\varphi}\right)^{m-n} \left(1 \mp i\frac{\rho'(\varphi)}{\rho(\varphi)}\right) \mathrm{d}\varphi = \delta_{mn}.$$

We preserve the second-order terms in formulas (5.11) and see that, for $m, n \gg 1$,

$$G_{mn} \cong \frac{1}{2}\delta_{mn} + b_{mn}, \quad G_{-m,-n} \cong \frac{1}{2}\delta_{mn} + b_{-m,-n}, \tag{5.12}$$

and

$$|b_{mn}|, |b_{-m,\,-n}| \le \frac{\text{const}}{|m-n|^{1/2}} \sigma_1^{m-n} \frac{(n-1)!}{(m+1)!}, \quad |m-n| \gg 1, \tag{5.13}$$

$$|b_{mn}|, |b_{-m,\,-n}| \le \text{const}/mn, \quad |m-n| \sim 1.$$

Thus, system (5.9) can be rewritten as

$$\frac{1}{2} a_n = a_n^0 + \sum_{m=-\infty}^{\infty} \left(G_{nm} - \frac{1}{2} \delta_{nm} \right) a_m. \tag{5.14}$$

Further, for $m \gg |n|, m > 0$, from (5.10) with (5.11) taken into account we obtain

$$G_{mn} \cong \frac{i^{m-n}}{4m!} \left(\frac{k}{2} \right)^m \int_0^{2\pi} \left[ik\,\rho(\varphi) H_n^{(2)\prime}(k\,\rho(\varphi)) + n \frac{\rho'(\varphi)}{\rho(\varphi)} H_n^{(2)}(k\,\rho(\varphi)) \right]$$
$$\times e^{in\varphi} (\rho(\varphi) e^{-i\varphi})^m d\varphi .$$

Proceeding as in Section 1.2.5, we obtain

$$|G_{mn}| \le \text{const} \frac{\sigma_1^m}{m!\,m^{1/2}},$$

where

$$\sigma_1 = \max_{\varphi_0, \alpha, s_\beta} Re\{F(\varphi_0)\}, \tag{5.15}$$

and (as in Section 1.2.5)

$$F(\varphi) = \frac{ik\,\rho(\varphi)}{2} e^{is_\beta(\varphi-\alpha)}, \quad s_\beta = \pm 1, 0 \le \alpha \le 2\pi, \tag{5.16}$$

where φ_0 are the roots of Equations (1.99) corresponding to the singular points of the mapping (1.55) that lie inside S [94].

If $|m| \gg |n|, m < 0$, then

$$|G_{mn}| \le \text{const} \frac{\sigma_1^{|m|}}{|m|!\,|m|^{1/2}}. \tag{5.17}$$

Similarly, for $|n| \gg |m|$, we can obtain

$$|G_{mn}| \le \text{const} \frac{|n|!}{\sigma_2^{|n|}\,|n|^{3/2}}, \tag{5.18}$$

where

$$\sigma_2 = \min_{\varphi_0, \alpha, s_\beta} \left| \frac{k\,\rho(\varphi_0)}{2} e^{is_\beta(\varphi_0-\alpha)} \right|, \tag{5.19}$$

and the minimum is taken over the set of roots φ_0 of equations (1.99) associated with points on the planes $\zeta^\pm$ which are now outside the contours $C^\pm$. Finally, for $n, m \gg 1$, but $|n - m| \sim 1$, we have

$$G_{-m,n} \cong \mathrm{const}\kappa(m+n), \quad G_{m,-n} \cong \mathrm{const}\overline{\kappa(m+n)}, \qquad (5.20)$$

where (the bar denotes the complex conjugation)

$$\kappa(m+n) = \int_0^{2\pi} (\rho(\varphi))^{m-n}\left(1 + i\frac{\rho'(\varphi)}{\rho(\varphi)}\right)e^{i(m+n)\varphi}\,d\varphi.$$

By the Riemann-Lebesgue theorem, the quantities $\kappa(m+n)$ decrease at a rate determined by the smoothness degree of the off-exponential term in the integrand. In particular, for $\rho(\varphi) \in C^\infty$, the quantities $\kappa(m+n)$ decrease faster than any power of $(m+n)$ [137].

The asymptotics of the coefficients a_n was obtained earlier (see Section 1.2.4), and it has the form

$$|a_n| < \mathrm{const}\frac{\sigma^{|n|}}{|n|!}, \qquad (5.21)$$

where (recall) $\sigma = kr_0/2$, and r_0 is the distance to the singularity of the mapping (1.55) that lies inside S and is most remote from the origin. For example, the value of σ is determined by formula (5.15) in the case of analytic boundary S and the initial field in the form of a plane wave.

Now we change the unknown variable in system (5.14) by the formula

$$a_n = \frac{|n|^{1/2}\sigma_1^{|n|}}{|n|!}x_n$$

and express a_0 in terms of the other unknowns to eliminate it from the system. Then system (5.14) becomes

$$x_n = x_n^0 + \sum_{m=-\infty}^{\infty}(1 - \delta_{m0})g_{nm}x_m, \quad n = \pm 1, \pm 2, \ldots, \qquad (5.22)$$

where

$$x_n^0 = 2\left(a_n^0 + \frac{G_{n0}a_0^0}{1 - G_{00}}\right)\frac{|n|!}{|n|^{1/2}\sigma_1^{|n|}},$$

$$g_{nm} = \left(2G_{nm} + 2\frac{G_{n0}G_{0m}}{1 - G_{00}} - \delta_{nm}\right)\sqrt{\left|\frac{m}{n}\right|}\frac{|n|!}{|m|!}\sigma_1^{|m|-|n|}.$$

It follows from the obtained estimates (5.12), (5.13), (5.17), (5.18), (5.20), and the easily verified inequality

$$|x_n^0| \leq \text{const} \frac{1}{n}$$

that, for

$$\sigma_2 \geq \sigma_1, \tag{5.23}$$

system (5.22) is Fredholm and can be solved (everywhere uniquely except for the set of spectrum points) by the reduction method (e.g., see [138]).

The scatterers whose boundaries satisfy condition (5.23) will be called weakly nonconvex [95]. Thus, the scatterers such that the singularities of the analytic deformation toward the interior of the domain lie closer to the origin than the singularities of the analytic deformation toward the outside are called weakly nonconvex scatterers.

Condition (5.23) is generally more restrictive than the condition under which the initial integral Equation (5.4) holds. Therefore, for the circle and ellipse, the system (5.22) is always solvable while for the multifoil (1.68), difficulties arise for $q = 2$. Even in the problem of plane wave scattering, where the set $\bar{B}_0$ is the interval

$$\left\{ |x| \leq \frac{2a}{3}(2 + \sqrt{1 + 3\tau^2}) \left(\frac{-1 + \sqrt{1 + 3\tau^2}}{3\tau} \right)^{1/2} \right\}$$

which is completely contained in the scatterer contour S, condition (5.23) leads to the following restriction on the "degree of nonconvexity": $\tau \leq 1/3$. For $q = 3$, system (5.22) is Fredholm for $\tau \leq 0,2214454...$, for $q = 4$, it is Fredholm for $\tau \leq 0,1659034...$, etc. We note that the considered figures are Rayleigh for $\tau < 0,2653...$ ($q = 2$), for $\tau < 0,1668...$ ($q = 3$), and for $\tau < 0,1216...$ ($q = 4$).

The results of numerical studies of the above-listed figures for $q = 3$ and $q = 4$ were confirmed by the fact that system (5.22) is Fredholm for τ, which does not exceed the above-listed threshold values.

5.2 WAVE DIFFRACTION AT A GROUP OF BODIES

As previously noted, when solving the diffraction problems where it is necessary to take into account the mutual influence of separate elements of the diffraction structure, PEM has a certain advantage over the other methods because the required quantity in PEM is a functional of the distribution of sources generating the diffraction field.

We illustrate the realization of this idea with an example of solving the diffraction problem at a group of ideally conducting cylindrical scatterers in the case where the electric vectors of the incident and scattered fields have a single component directed along the generating elements of these cylindrical bodies. Then the problem becomes two-dimensional, and the Dirichlet boundary condition is satisfied on the boundaries of the cylindrical bodies. The problem geometry is shown in Fig. 5.1.

We write the system of PEM integral-operator equations for the problem of diffraction at N bodies [98,99]

$$g_j(\alpha) = g_j^0(\alpha) + \frac{1}{\pi}\int_0^{2\pi}\int_\Gamma \frac{k}{4}\left[\rho_j\left(\varphi_j\right)\cos\psi - \rho_j'\left(\varphi_j\right)\sin\psi\right]g_j\left(\varphi_j+\psi\right)$$
$$\times \exp\left\{-ik\rho_j\left(\varphi_j\right)\left[\cos\psi - \cos\left(\alpha-\varphi_j\right)\right]\right\}d\psi d\varphi_j$$
$$+ \sum_{l=1}^N\left(1-\delta_{jl}\right)\frac{1}{\pi}\int_0^{2\pi}\exp\left\{ik\rho_j\left(\varphi_j\right)\cos\left(\alpha-\varphi_j\right)\right\}$$
$$D_j\int_\Gamma g_l(\varphi_l+\psi)e^{-ik\eta\cos\psi}d\psi d\varphi_{jj} = \overline{1,N}. \tag{5.24}$$

In this system, we have $g_j(\varphi)$ is the scattering pattern of the jth body [98,99],

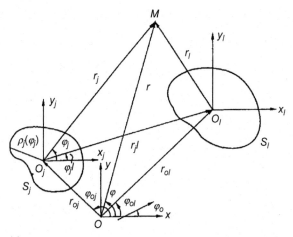

Figure 5.1 Problem geometry.

$$g_j^0(\alpha) = \frac{k}{4} e^{-i\vec{k}\vec{r}_{0j}} \int_0^{2\pi} \left[\rho_j(\varphi_j) \cos(\varphi_j - \varphi_0) + \rho_j'(\varphi_j) \sin(\varphi_j - \varphi_0) \right]$$

$$\times \exp\left\{ ik\,\rho_j(\varphi_j) \left[\cos(\alpha - \varphi_j) - \cos(\varphi_0 - \varphi_j) \right] \right\} d\varphi_j. \quad (5.25)$$

r_j, φ_j are polar coordinates of the observation point at the center of the jth body, $r_j = \rho_j(\varphi_j)$ is the equation of the contour S_j of the jth body cross-section, $\vec{k}\,\vec{r}_{0j} = kr_{0j}\cos(\varphi_0 - \varphi_{0j})$, and r_{0j}, φ_{0j} are the coordinates of the jth body center with respect to the common origin, φ_0 is the angle of incidence of the initial plane wave, and $D_j = \frac{i}{4}\left[\rho_j(\varphi_j)\frac{\partial}{\partial r_j} - \frac{\rho_j'(\varphi_j)}{\rho_j(\varphi_j)}\frac{\partial}{\partial \varphi_j} \right]\big|_{r_j = \rho_j(\varphi_j)}$ is the differentiation operator.

We use the expansions in the Fourier series

$$g_j(\alpha) = \sum_{m=-\infty}^{\infty} c_{jm} e^{im\alpha}, \quad g_j^0(\alpha) = \sum_{m=-\infty}^{\infty} c_{jm}^0 e^{im\alpha}, \quad j = \overline{1, N} \quad (5.26)$$

to obtain from (5.24) the following system of algebraic equations for the coefficients c_{jm}:

$$c_{jm} = c_{jm}^0 + \sum_{l=1}^{N} \sum_{n=-\infty}^{\infty} G_{mn}^{jl} c_{ln}, \quad j = \overline{1, N}, m = 0, \pm 1, \dots \quad (5.27)$$

In this system, we have

$$G_{mn}^{jj} = \frac{i^{m-n+1}}{4} \int_0^{2\pi} J_m(k\rho_j) \left[k\rho_j(\varphi_j) H_n^{(2)\prime}(k\rho_j) - in\frac{\rho_j'(\varphi_j)}{\rho_j(\varphi_j)} H_n^{(2)}(k\rho_j) \right]$$

$$\times e^{i(n-m)\varphi_j} d\varphi_j$$

$$(5.28)$$

$$G_{mn}^{jl} = \frac{i^{m-n+1}}{4} \sum_{p=-\infty}^{\infty} H_{n-p}^{(2)}(kr_{lj}) e^{i(n-p)\varphi_{lj}} \int_0^{2\pi} J_m(k\rho_j) \left[k\rho_j(\varphi_j) J_p'(k\rho_j) \right.$$

$$\left. - ip\frac{\rho_j'(\varphi_j)}{\rho_j(\varphi_j)} J_p(k\rho_j) \right] e^{i(p-m)\varphi_j} d\varphi_j, \quad j \neq l$$

$$(5.29)$$

$$c_{jm}^0(\varphi_0) = e^{-i\vec{k}\vec{r}_{0j}} \frac{i^{m+1}}{4} \sum_{p=-\infty}^{\infty} (-i)^p e^{-ip\varphi_0} \int_0^{2\pi} J_m\left(k\rho_j\right) \left[k\rho_j\left(\varphi_j\right)J_p'\left(k\rho_j\right)\right.$$

$$\left. -ip\frac{\rho_j'\left(\varphi_j\right)}{\rho_j\left(\varphi_j\right)}J_p\left(k\rho_j\right)\right] e^{i(p-m)\varphi_j} d\varphi_j \equiv e^{-i\vec{k}\vec{r}_{0j}} c_{j0m}^0(\varphi_0) \tag{5.30}$$

Relation (5.30) allows us to rewrite formula (5.29) as

$$G_{mn}^{jl} = e^{in\varphi_{lj}} \frac{1}{\pi}\int_\Gamma \exp\left(-ikr_{lj}\cos\psi + in\psi\right) c_{j0m}^0\left(\psi + \varphi_{lj}\right) d\psi. \tag{5.31}$$

If we use (5.31), then system (5.27) becomes

$$c_{jm} - \sum_{n=-\infty}^{\infty} G_{mn}^{jj} c_{jn} = e^{-i\vec{k}\vec{r}_{0j}} c_{j0m}^0 + \sum_{l=1}^{N}(1-\delta_{jl})\frac{1}{\pi}\int_\Gamma e^{-ikr_{jl}\cos\psi} g_l\left(\varphi_{lj}+\psi\right)$$

$$\times c_{j0m}^0\left(\varphi_{lj}+\psi\right) d\psi \tag{5.32}$$

For $kr_{lj}\gg 1$, we can obtain the asymptotics solution of the problem. For this, we rewrite system (5.32) in the vector-matrix notation as

$$G^{jj}\vec{c}_j(\varphi_0) = e^{-i\vec{k}\vec{r}_{0j}} \vec{c}_{j0}^0(\varphi_0) + \sum_{l=1}^{N}(1-\delta_{jl})\frac{1}{\pi}\int_\Gamma e^{-ikr_{jl}\cos\psi} g_l\left(\varphi_{lj}+\psi;\varphi_0\right)$$

$$\times \vec{c}_{j0}^0\left(\varphi_{lj}+\psi\right) d\psi \tag{5.33}$$

where

$$\left(G^{jj}\vec{c}_j\right)_m = c_{jm} - \sum_{n=-\infty}^{\infty} G_{mn}^{jj} c_{jn}.$$

For $kr_{lj}=\infty$, we have

$$G^{jj}\vec{c}_j^\infty(\varphi_0) = \vec{c}_{j0}^0(\varphi_0), \text{ i.e., } \left(G^{jj}\right)^{-1}\vec{c}_{j0}^0(\varphi_0) = \vec{c}_j^\infty(\varphi_0),$$

and hence (5.33) implies

$$c_{jm}(\varphi_0) = e^{-i\vec{k}\vec{r}_{0j}} c_{jm}^\infty(\varphi_0) + \sum_{l=1}^{N}(1-\delta_{jl})\frac{1}{\pi}\int_\Gamma e^{-ikr_{jl}\cos\psi} g_l\left(\varphi_{lj}+\psi;\varphi_0\right) c_{jm}^\infty\left(\varphi_{lj}+\psi\right) d\psi. \tag{5.34}$$

For $kr_{lj} \gg 1$, estimating the integral in (5.34) by the saddle point method, we obtain

$$c_{jm}(\varphi_0) \cong e^{-i\vec{k}\vec{r}_{0j}} c_{jm}^{\infty}(\varphi_0) + \sum_{l=1}^{N}(1-\delta_{jl}) Q_{jl} g_l\left(\varphi_{lj}; \varphi_0\right) c_{jm}^{\infty}\left(\varphi_{lj}\right),$$

where $Q_{jl} = \sqrt{\dfrac{2}{\pi k r_{lj}}} e^{-ikr_{lj} + i\pi/4}$.

We multiply the obtained asymptotic relation by $e^{in\varphi}$ and calculate the sum. Then we finally obtain

$$g_j(\varphi; \varphi_0) \cong e^{-i\vec{k}\vec{r}_{0j}} g_j^{\infty}(\varphi; \varphi_0) + \sum_{l=1}^{N}(1-\delta_{jl}) Q_{jl} g_j^{\infty}\left(\varphi; \varphi_{lj}\right) g_l\left(\varphi_{lj}; \varphi_0\right),$$

$$(5.35)$$

where $g_j^{\infty}(\varphi; \varphi_0)$ is the scattering pattern of the jth body without the influence of other bodies taken into account.

System (5.35) allows one easily to obtain an approximate solution of the problem. In particular, for scatterers of sufficiently small dimensions (for $\max_j d_j < \lambda/8$, where d_j is the maximal dimension of the jth body), we can determine the desired quantities $g_j(\varphi; \varphi_0)$ with an appropriate accuracy up to the contact of separate scatterers. The coefficients $g_l(\varphi_{lj}; \varphi_0)$ required in calculations can be determined from the algebraic systems

$$g_j\left(\varphi_{jq}; \varphi_0\right) - \sum_{l=1}^{N}(1-\delta_{jl}) Q_{jl} g_j^{\infty}\left(\varphi_{jq}; \varphi_{lj}\right) g_l\left(\varphi_{lj}; \varphi_0\right) = e^{-i\vec{k}\vec{r}_{0j}} g_j^{\infty}\left(\varphi_{jq}; \varphi_0\right),$$

$$j = \overline{1, N}, \quad q = \overline{1, N}. \tag{5.36}$$

In the case where the scatterers are circular cylinders, relations (5.28)–(5.30) become

$$G_{mn}^{jj} = \frac{i\pi k a_j}{2} J_m(ka_j) H_m^{(2)\prime}(ka_j)\delta_{mn}, \tag{5.37}$$

$$G_{mn}^{jl} = \frac{i\pi k a_j}{2} J_m(ka_j) J_m'(ka_j) H_{n-m}^{(2)}(kr_{lj}) e^{i(n-m)(\varphi_{lj} - \pi/2)}, \quad j \neq l, \tag{5.38}$$

$$c_{jm}^{0}(\varphi_0) = e^{-i\vec{k}\vec{r}_{0j}} \frac{i\pi k a_j}{2} J_m(ka_j) J_m'(ka_j) e^{-im\varphi_0}, \tag{5.39}$$

where a_j are the cylinder radii.

If $ka_j \ll 1$, then it follows from (5.37)–(5.39) that algebraic system (5.27) in the one-mode approximation becomes

$$c_{j0}H_0^{(2)}(ka_j) + J_0(ka_j)\sum_{l=1}^{N}(1-\delta_{jl})H_0^{(2)}(kr_{lj})c_{l0} = -\mathrm{e}^{-i\vec{k}\vec{r}_{0j}}J_0(ka_j), \quad j=\overline{1,N}.$$

(5.40)

Relations (5.40) underlie the idea of the method of elementary scatterers (MES) [130,131], which consists in that the scatterer of an arbitrary geometry and structure can be replaced by a group of bodies of the same type (for example, circular cylinders) which in the whole "repeat" the geometry and structure of the initial scatterer.

Let us study the asymptotics of matrix elements and constant terms in system (5.27) as $|m|, |n| \to \infty$. The asymptotics of the elements G_{mn}^{ij} was in fact studied above (see Section 5.1.2, relations (5.12)–(5.20) up to the replacement of σ_1 by $\sigma_1^{(j)}$ and σ_2 by $\sigma_2^{(j)}$).

For the elements G_{mn}^{jl}, the estimates for $|m| \gg |n|$ can be obtained by analogy with the estimates for the elements G_{mn}^{ij}. As a result, we obtain

$$|G_{mn}^{jl}| \le \mathrm{const}\frac{\sigma_1^{(j)|m|}}{|m|!|m|^{1/2}}, \quad j \ne l.$$

(5.41)

For $|n| \gg |m|$, using the asymptotic relations (5.11), we derive from (5.29)

$$G_{mn}^{jl} \cong \frac{i^{m-n+1}}{4\pi}|n|!\left(\frac{2}{k}\right)^{|n|}\int_0^{2\pi}J_m(k\rho_j)\mathrm{e}^{-im\varphi_j}\frac{\left[\rho_j'(\varphi_j) - is_n\rho_j(\varphi_j)\right]\mathrm{e}^{-is_n\varphi_j}}{r_{lj}\mathrm{e}^{-is_n\varphi_{lj}} + \rho_j(\varphi_j)\mathrm{e}^{-is_n\varphi_j}}$$

$$\mathrm{e}^{-|n|\Phi_{jn}(\varphi_j)}\mathrm{d}\varphi_j,$$

(5.42)

where

$$\Phi_{jn}(\varphi_j) = \ln\left[r_{lj}\mathrm{e}^{-is_n\varphi_{lj}} + \rho_j(\varphi_j)\mathrm{e}^{-is_n\varphi_j}\right], \quad s_n = n/|n|.$$

We let α_{mn} denote the numerical multiplier before the integral in (5.42) and pass to the new variable in the integral:

$$\zeta = \rho_j(\varphi_j)\mathrm{e}^{-is_n\varphi_j}.$$

(5.43)

Then relation (5.42) becomes

$$G_{mn}^{jl} \cong \alpha_{mn}\int_{C_j}\frac{f_m(\zeta)\mathrm{d}\zeta}{\left(r_{lj}\mathrm{e}^{-is_n\varphi_{lj}} + \zeta\right)^{|n|+1}},$$

(5.44)

where $f_m(\zeta)$ denotes the function $J_m\left(k\rho_j\right)e^{-im\varphi_j}$ after the change (5.43). The contour C_j coincides with S_j for $s_n = -1$ and is the mirror reflection of S_j with respect to the axis Ox for $s_n = +1$ on the complex plane z. The estimate

$$|G_{mn}^{jl}| \le |\alpha_{mn}|\int_{C_j}\left|\frac{f_m(\zeta)d\zeta}{\left(r_{lj}e^{-is_n\varphi_{lj}} + \zeta\right)^{|n|+1}}\right|$$

is exact for the minimal value on the right-hand side, and since $\left|r_{lj}e^{-is_n\varphi_{lj}} + \zeta\right|$ is the distance from O_l to a point on the contour C_j, it is necessary to contract the contour C_j so as to obtain the maximal value of this quantity. To obtain an exact estimate of the integral in (5.42), we use the saddle point method with the some considerations about the direction of deformation of the contour C_j. As a result, we obtain the estimate

$$|G_{mn}^{jl}| \le \frac{\text{const}|n|!}{\sigma_{jl}^{|n|}|n|^{3/2}}, \quad |n|\gg|m|, \quad j\ne l, \tag{5.45}$$

and

$$\sigma_{jl} = \frac{1}{2}\min_{\varphi_{j0},s}\left|kr_{lj}e^{-is\varphi_{lj}} + k\rho_j\left(\varphi_{j0}\right)e^{-is\varphi_{j0}}\right|, \quad j\ne l, \tag{5.46}$$

where φ_{j0} are the roots of equations (1.99) associated with points inside the contours C_j. Thus, $2\sigma_{jl}/k$ is equal to the distance from O_l to the nearest singularity of the analytic deformation of the contour S_j toward the interior.

For the constant terms, proceeding as previously, we obtain

$$|a_{jm}^0| \le \text{const}\frac{\sigma_j^{|m|}}{|m|!|m|^{1/2}}, \tag{5.47}$$

where

$$\sigma_j = \max\left\{\sigma_1^{(j)}, \sigma_0^{(j)}\right\},$$

and $\sigma_0^{(j)} = kr_0^{(j)}/2$, $r_0^{(j)}$ is the distance to the point inside S_j that is most remote from O_j and corresponds to a singularity of the function $v_0^{(j)}(\varphi_j)$ (i.e., to the boundary value of the incident field) as the function is continued into the domain of complex angles. Since the functions $v_0^{(j)}(\varphi_j)$ are explicitly known, they can easily be found (see Chapter 1).

Now we eliminate the coefficients c_{j0} from system (5.27) and replace the unknown coefficients by the formula

$$c_{jm} = \frac{|m|^{1/2}\sigma_j^{|m|}}{|m|!}x_{jm}, \quad j = \overline{1,N}, m = \pm 1, \pm 2, \dots.$$

As a result, we obtain a new system for the coefficients x_{jm}:

$$x_{jm} = x_{jm}^0 + \sum_{n=-\infty}^{\infty}\sum_{l=1}^{N}(1-\delta_{n0})g_{mn}^{jl}x_{ln}, \quad j = \overline{1,N}, m = \pm 1, \pm 2, \dots, \quad (5.48)$$

with

$$x_{jm}^0 = \frac{|m|!\hat{c}_{jm}^0}{|m|^{1/2}\sigma_j^{|m|}}, \quad g_{mn}^{ij} = \left|\frac{n}{m}\right|^{1/2}\frac{|m|!}{|n|!}\sigma_j^{|n|-|m|}\hat{G}_{mn}^{ij},$$

$$g_{mn}^{jl} = \left|\frac{n}{m}\right|^{1/2}\frac{|m|!\sigma_l^{|n|}}{|n|!\sigma_j^{|m|}}\hat{G}_{mn}^{jl}, \quad j \neq l,$$

where $\hat{c}_{jm}^0, \hat{G}_{mn}^{ij}, \hat{G}_{mn}^{jl}$ can obviously be expressed in terms of $\hat{c}_{jm}^0, G_{mn}^{ij}, G_{mn}^{jl}$. Thus, for $N = 2$, we have [98]

$$\hat{c}_{jm}^0 = c_{jm} + \frac{1}{\Delta}\left\{G_{m0}^{j1}\left[G_{00}^{12}c_{20}^0 + \left(1-G_{00}^{22}\right)c_{10}^0\right] + G_{m0}^{j2}\left[G_{00}^{21}c_{10}^0 + \left(1-G_{00}^{11}\right)c_{20}^0\right]\right\},$$

$$\hat{G}_{mn}^{ij} = G_{mn}^{ij} + \frac{1}{\Delta}\left[G_{m0}^{ij}G_{00}^{ji}G_{0n}^{ij} + G_{m0}^{ji}G_{00}^{ij}G_{0n}^{ij} + G_{m0}^{ij}\left(1-G_{00}^{ij}\right)G_{0n}^{ij} + G_{m0}^{ji}\left(1-G_{00}^{ij}\right)G_{0n}^{ij}\right],$$

$$\hat{G}_{mn}^{ji} = G_{mn}^{ji} + \frac{1}{\Delta}\left[G_{m0}^{ji}G_{00}^{ij}G_{0n}^{ji} + G_{m0}^{ij}G_{00}^{ji}G_{0n}^{ii} + G_{m0}^{ij}\left(1-G_{00}^{ii}\right)G_{0n}^{ij} + G_{m0}^{ji}\left(1-G_{00}^{ij}\right)G_{0n}^{ii}\right], i \neq j,$$

$$\Delta = \left(1-G_{00}^{11}\right)\left(1-G_{00}^{22}\right) - G_{00}^{12}G_{00}^{21}.$$

As an example, we consider the problem of diffraction at two bodies by using the rigorous approach (relations (5.27)–(5.30)) and the asymptotic solution (5.35). Figures 5.2 and 5.3, respectively, show the rigorous and approximate solutions of the problem of diffraction at two circular cylinders of radius $ka = 3$ located at the distances $kl = 15$ and $kl = 45$ between the cylinder centers. The initial wave propagates at the angle $\varphi_0 = \pi/2$ to the line connecting the cylinder centers (axis Ox). The calculations were performed for the maximal harmonic number M in (5.26), which is equal to 6. Digit 1 denotes the rigorous solution of the problem, and digit 2 denotes the asymptotic solution. One can see that the curves practically coincide for $kl = 45$. The ratio $l/2a$ is here equal to 7.5.

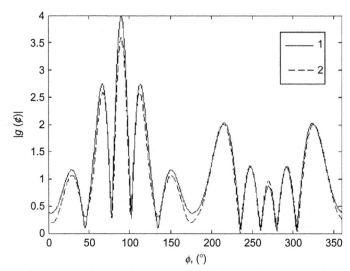

Figure 5.2 Rigorous and approximate solutions of the diffraction problem at two circular cylinders located at the distance $kl = 15$.

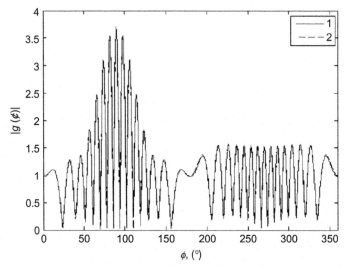

Figure 5.3 Rigorous and approximate solutions of the diffraction problem at two circular cylinders located at the distance $kl = 45$.

Figure 5.4 illustrates the results of similar calculations in the case of two elliptic cylinders with semiaxes $ka = 5$ (along the axis Ox) and $kb = 3$ where the distance between the cylinder centers is $kl = 40$. The maximal harmonic number M is here equal to 12.

In the case under study, the ratio $l/2a$ is equal to 4, i.e., it is almost two times less than that in the case illustrated in Fig. 5.3, which explains

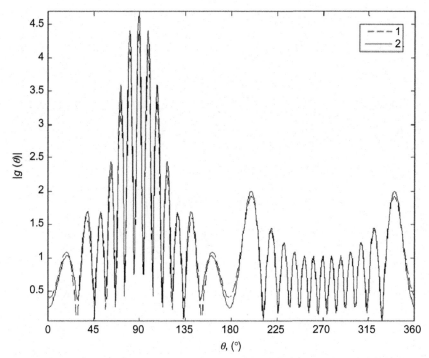

Figure 5.4 Rigorous and approximate solutions of the diffraction problem at two elliptic cylinders with semiaxes $ka = 5$ (along the axis Ox) and $kb = 3$, the distance between the cylinder centers is $kl = 40$.

the noticeably large difference between the rigorous and asymptotic solutions.

As the second example, we consider the realization of the method of elementary scatterers applied to the problem of diffraction at thin screens [131]. We consider the diffraction of a plane wave incident along the axis of symmetry on a screen shaped as a cylinder of radius $ka = 15$ with the opening angle $90°$. In Fig. 5.5, digit 1 indicates the scattering pattern of a grating of $N = 20$ circular cylinders of radius $ka = 0.25$ (the maximal possible number of cylinders of such a radius, which can be located on the considered arc, is $N = 46$) located along the screen directrix. Digit 2 in this and subsequent figures indicates the scattering pattern of the considered screen obtained by the method of continued boundary conditions (see Chapter 4).

One can see that there is a rather noticeable distinction between the curves. Figure 5.6 shows the result of calculations for $ka = 0.025$ and $N = 400$ (the maximal possible number of cylinders is $N = 460$ in this case). The total graphical coincidence with the exact result is obvious.

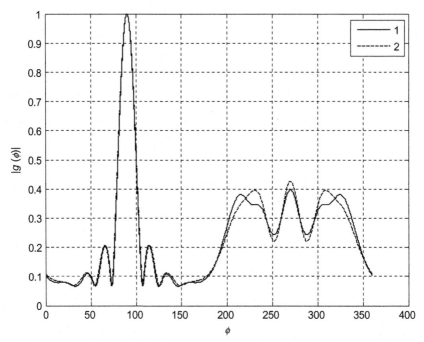

Figure 5.5 Method of elementary scatterers at $ka = 0.25$ and $N = 20$.

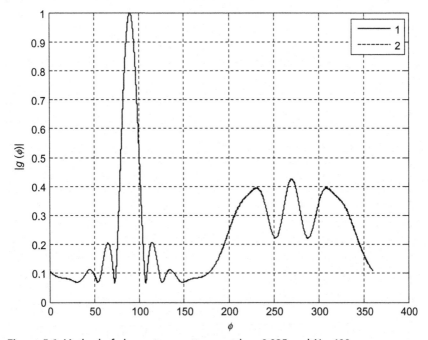

Figure 5.6 Method of elementary scatterers at $ka = 0.025$ and $N = 400$.

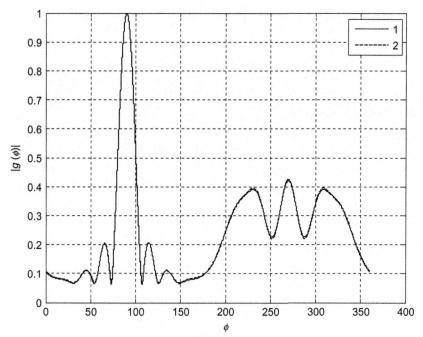

Figure 5.7 Method of elementary scatterers at $ka = 0.025$ and $N = 100$.

Finally, Fig. 5.7 illustrates the result of similar calculations for $N = 100$. One can see that, even in this case, the result is quite acceptable, despite the fact that the number of cylinders is 4.6 times less than in the case where the screen arc is filled completely. Only for $N = 50$, there is a slight graphical distinction between the results. Obviously, the picture for screens of any arbitrary geometry is similar. We note that the size of the algebraic system sufficient for the graphical accuracy is of the same order as in the method of singular integral equations [11] ($N \sim (10 \div 15)L/\lambda$, where L is the length of the scatterer boundary) but the matrix elements of the system in our case are of a significantly simper form.

Thus, the considered examples allow us to conclude that the proposed approach allows one efficiently (i.e., with minimal expenditure of computational resources) and with an acceptable accuracy to model the scattering characteristics of bodies of arbitrary geometry. The approach can easily be generalized to problems of modeling the scattering characteristics for impedance bodies and arbitrary inhomogeneous bodies, where it can be treated as an alternative to the method of volume integral equations.

5.3 WAVE DIFFRACTION AT PERIODIC GRATINGS

In this section, we consider the problem of wave diffraction at a multirow grating.

So let a plane wave be incident at an angle φ_0 to the axis OX on an M-row periodic grating (the elements in different rows, as well as the periods, can be different) (see Fig. 5.8).

By analogy with Kyurkchan [44], we can write the system of integral-operator equations of the second type with respect to the pattern $g_{0t}(\alpha; \varphi_0)$ of the central element in the tth ($t = 1, \ldots, M$) row of the grating. We assume that the Dirichlet boundary condition is satisfied on the grating elements. Thus, this system of equations has the form

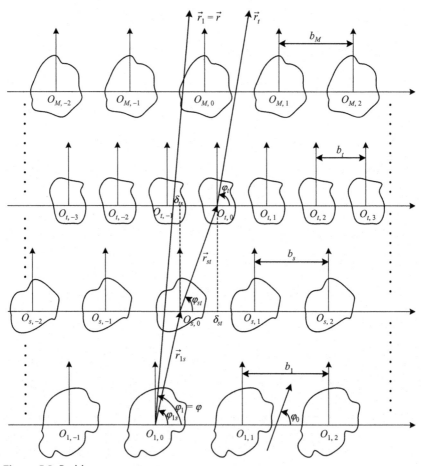

Figure 5.8 Problem geometry.

$$g_{0t}(\alpha; \varphi_0) = \exp[-ikr_{1t}\cos(\varphi_{1t} - \varphi_0)]\tilde{g}_{0t}^0(\alpha; \varphi_0)$$

$$+ \frac{1}{\pi}\int_0^{2\pi} \int_{-(\pi/2)-i\infty}^{\pi/2+i\infty} \frac{k}{4}[\rho_t(\varphi_t)\cos\psi - \rho_t'(\varphi_t)\sin\psi]g_{0t}(\varphi_t + \psi; \varphi_0)$$

$$\times \exp\{-ik\rho_t(\varphi_t)[\cos\psi - \cos(\alpha - \varphi_t)]\}d\psi d\varphi_t$$

$$+ \sum_{j=1}^{\infty}\left[e^{-ijkb_t\cos\varphi_0}\frac{1}{\pi}\int_{-(\pi/2)-i\infty}^{\pi/2+i\infty} e^{-ijkb_t\cos\psi}\tilde{g}_{0t}^0(\alpha; \psi - \pi)g_{0t}(\psi - \pi; \varphi_0)d\psi \right.$$

$$\left. + e^{ijkb_t\cos\varphi_0}\frac{1}{\pi}\int_{-(\pi/2)-i\infty}^{\pi/2+i\infty} e^{-ijkb_t\cos\psi}\tilde{g}_{0t}^0(\alpha; \psi)g_{0t}(\psi; \varphi_0)d\psi \right]$$

$$+ \sum_{s=1}^{t-1}\sum_{j=-\infty}^{\infty} e^{-ijkb_s\cos\varphi_0}\frac{1}{\pi}\int_{-i\infty}^{\pi+i\infty} e^{ijkb_s\cos\psi}e^{-ikr_{st}\cos(\psi-\varphi_{st})}\tilde{g}_{0t}^0(\alpha; \psi)g_{0s}(\psi; \varphi_0)d\psi$$

$$+ \sum_{s=t+1}^{M}\sum_{j=-\infty}^{\infty} e^{-ijkb_s\cos\varphi_0}\frac{1}{\pi}\int_{-i\infty}^{\pi+i\infty} e^{ijkb_s\cos\psi}e^{-ikr_{st}\cos(\psi+\varphi_{st})}$$

$$\tilde{g}_{0t}^0(\alpha; -\psi)g_{0s}(-\psi; \varphi_0)d\psi, \quad t = 1, ..., M. \tag{5.49}$$

In this system,

$$\tilde{g}_{0t}^0(\alpha; \varphi_0) = \frac{k}{4}\int_0^{2\pi} [\rho_t(\varphi_t)\cos(\varphi_t - \varphi_0) + \rho_t'(\varphi_t)\sin(\varphi_t - \varphi_0)]$$

$$e^{ik\rho_t(\varphi_t)[\cos(\alpha-\varphi_t)-\cos(\varphi_t-\varphi_0)]}d\varphi_t,$$

$r = \rho_t(\varphi_t)$ is the equation of the boundary of the central element in the tth row in local coordinates, $\vec{r}_{st} = \{r_{st}, \varphi_{st}\}$ is the radius vector from the origin in the central element in the sth row to the origin in the central element in the tth row (see Fig. 5.8), and b_t is the period of the tth row in the grating.

We can rewrite the system of Equations (5.49) as follows:

$$L_t[g_{0t}(\alpha; \varphi_0)] = e^{-ikr_{1t}\cos(\varphi_{1t}-\varphi_0)}\tilde{g}_{0t}^0(\alpha; \varphi_0)$$

$$+ \sum_{j=1}^{\infty}\left[e^{-ijkb_t\cos\varphi_0}\frac{1}{\pi}\int_{-(\pi/2)-i\infty}^{\pi/2+i\infty} e^{-ijkb_t\cos\psi}\tilde{g}_{0t}^0(\alpha; \psi - \pi)g_{0t}(\psi - \pi; \varphi_0)d\psi \right.$$

$$\left. + e^{ijkb_t\cos\varphi_0}\frac{1}{\pi}\int_{-(\pi/2)-i\infty}^{\pi/2+i\infty} e^{-ijkb_t\cos\psi}\tilde{g}_{0t}^0(\alpha; \psi)g_{0t}(\psi; \varphi_0)d\psi \right]$$

$$+ \sum_{s=1}^{t-1}\sum_{j=-\infty}^{\infty} e^{-ijkb_s\cos\varphi_0}\frac{1}{\pi}\int_{-i\infty}^{\pi+i\infty} e^{ijkb_s\cos\psi}e^{-ikr_{st}\cos(\psi-\varphi_{st})}\tilde{g}_{0t}^0(\alpha; \psi)g_{0s}(\psi; \varphi_0)d\psi$$

$$+ \sum_{s=t+1}^{M}\sum_{j=-\infty}^{\infty} e^{-ijkb_s\cos\varphi_0}\frac{1}{\pi}\int_{-i\infty}^{\pi+i\infty} e^{ijkb_s\cos\psi}e^{-ikr_{st}\cos(\psi+\varphi_{st})}\tilde{g}_{0t}^0(\alpha; -\psi)$$

$$g_{0s}(-\psi; \varphi_0)d\psi, \quad t = 1, ..., M.$$

Since

$$L_t\left[g_{0t}^\infty(\alpha;\varphi_0)\right]=\tilde{g}_{0t}^0(\alpha;\varphi_0),\quad t=1,\dots,M,$$

(see Section 5.1), where $g_{0t}^\infty(\alpha;\varphi_0)$ is the scattering pattern of the central element in the tth row in the grating under the assumption that all other grating elements are moved to infinity, the linearity of the operators L_t then implies that the functions $g_{0t}(\alpha;\varphi_0)$ satisfy the system of integral equations of the second type

$$g_{0t}(\alpha;\varphi_0)=e^{-ikr_{1t}\cos(\varphi_{1t}-\varphi_0)}g_{0t}^\infty(\alpha;\varphi_0)$$

$$+\sum_{j=1}^\infty\left[e^{-ijkb_t\cos\varphi_0}\frac{1}{\pi}\int_{-(\pi/2)-i\infty}^{\pi/2+i\infty}e^{-ijkb_t\cos\psi}g_{0t}^\infty(\alpha;\psi-\pi)g_{0t}(\psi-\pi;\varphi_0)d\psi\right.$$

$$\left.+e^{ijkb_t\cos\varphi_0}\frac{1}{\pi}\int_{-(\pi/2)-i\infty}^{\pi/2+i\infty}e^{-ijkb_t\cos\psi}g_{0t}^\infty(\alpha;\psi)g_{0t}(\psi;\varphi_0)d\psi\right]$$

$$+\sum_{s=1}^{t-1}\sum_{j=-\infty}^\infty e^{-ijkb_s\cos\varphi_0}\frac{1}{\pi}\int_{-i\infty}^{\pi+i\infty}e^{ijkb_s\cos\psi}e^{-ikr_{st}\cos(\psi-\varphi_{st})}g_{0t}^\infty(\alpha;\psi)g_{0s}(\psi;\varphi_0)d\psi$$

$$+\sum_{s=t+1}^M\sum_{j=-\infty}^\infty e^{-ijkb_s\cos\varphi_0}\frac{1}{\pi}\int_{-i\infty}^{\pi+i\infty}e^{ijkb_s\cos\psi}e^{-ikr_{st}\cos(\psi+\varphi_{st})}$$

$$g_{0t}^\infty(\alpha;-\psi)g_{0s}(-\psi;\varphi_0)d\psi,\quad t=1,\dots,M. \tag{5.50}$$

Such equations were first obtained by Tversky (e.g., see [139]). The principal distinction of the system of Equations (5.50) from system (5.49) is that to solve the problem of diffraction at a grating by using (5.50), it is first necessary to solve the problem of a plane wave scattering at a solitary element in each row of the grating (i.e., to find the functions $g_{0t}^\infty(\alpha;\varphi_0)$ for $t=1,\dots,M$), while system (5.49) immediately allows one to find the pattern of any element in the grating. However, the system of Equations (5.50) is convenient when it is required to obtain an asymptotic (see below) solution of the problem.

For $kb_t\gg1$, $t=1,\dots,M$, and for $kr_{st}\gg1$, $s,t=1,\dots,M,s\ne t$, the integrals in (5.50) can be calculated by the saddle point method. As a result, we obtain the following system of M algebraic equations for the functions $g_{0t}(\alpha;\varphi_0)$:

$$g_{0t}(\alpha; \varphi_0) \cong e^{-ikr_{1t}\cos(\varphi_{1t}-\varphi_0)}g_{0t}^\infty(\alpha; \varphi_0)$$

$$+ \sum_{j=1}^\infty \sqrt{\frac{2}{\pi jkb_t}} e^{-ijkb_t + i\pi/4}\left[e^{-ijkb_t\cos\varphi_0}g_{0t}^\infty(\alpha; \pi)g_{0t}(\pi; \varphi_0)\right.$$

$$\left. + e^{ijkb_t\cos\varphi_0}g_{0t}^\infty(\alpha; 0)g_{0t}(0; \varphi_0)\right] + \sum_{s=1}^{t-1}\left\{\sqrt{\frac{2}{\pi kr_{st}}}e^{-ikr_{st} + i\pi/4}g_{0t}^\infty(\alpha; \varphi_{st})g_{0s}(\varphi_{st}; \varphi_0)\right.$$

$$+ \sum_{j=1}^\infty \sqrt{\frac{2}{\pi jkb_s}} e^{-ijkb_s + i\pi/4}\left[e^{-ijkb_s\cos\varphi_0}e^{ikr_{st}\cos\varphi_{st}}g_{0t}^\infty(\alpha; \pi)g_{0s}(\pi; \varphi_0)\right.$$

$$\left.\left. + e^{ijkb_s\cos\varphi_0}e^{-ikr_{st}\cos\varphi_{st}}g_{0t}^\infty(\alpha; 0)g_{0s}(0; \varphi_0)\right]\right\}$$

$$+ \sum_{s=t+1}^M \left\{\sqrt{\frac{2}{\pi kr_{st}}}e^{-ikr_{st} + i\pi/4}g_{0t}^\infty(\alpha;\varphi_{st} - \pi)g_{0s}(\varphi_{st} - \pi;\varphi_0)\right.$$

$$+ \sum_{j=1}^\infty \sqrt{\frac{2}{\pi jkb_s}} e^{-ijkb_s + i\pi/4}\left[e^{-ijkb_s\cos\varphi_0}e^{ikr_{st}\cos\varphi_{st}}g_{0t}^\infty(\alpha; \pi)g_{0s}(\pi; \varphi_0)\right.$$

$$\left.\left. + e^{ijkb_s\cos\varphi_0}e^{-ikr_{st}\cos\varphi_{st}}g_{0t}^\infty(\alpha; 0)g_{0s}(0; \varphi_0)\right]\right\}.$$

$$(5.51)$$

This system of equations allows one approximately to find $g_{0t}(\alpha;\varphi_0)$ under the condition that the periods of the grating rows and the distances between the rows are sufficiently large. Below we present the results of calculations by formulas (5.51).

To find the rigorous numerical solution of the problem, we perform its algebraization. Substituting the expansions in the Fourier series

$$g_{0s}(\alpha; \varphi_0) = \sum_{m=-\infty}^\infty a_{sm}e^{ima}, \quad e^{-ikr_{1s}\cos(\varphi_{1s}-\varphi_0)}\widetilde{g}_{0s}^0(\alpha; \varphi_0) = \sum_{m=-\infty}^\infty a_{sm}^0 e^{ima}$$

into (5.49), we obtain the system of PEM algebraic equations

$$a_{1m} = a_{1m}^0 + \sum_{n=-\infty}^\infty \left(G_{mn}^{11}a_{1n} + \cdots + G_{mn}^{1t}a_{tn} + \cdots + G_{mn}^{1M}a_{Mn}\right),$$

$$a_{sm} = a_{sm}^0 + \sum_{n=-\infty}^\infty \left(G_{mn}^{s1}a_{1n} + \cdots + G_{mn}^{st}a_{tn} + \cdots + G_{mn}^{sM}a_{Mn}\right), \qquad (5.52)$$

$$a_{Mm} = a_{Mm}^0 + \sum_{n=-\infty}^\infty \left(G_{mn}^{M1}a_{1n} + \cdots + G_{mn}^{Mt}a_{tn} + \cdots + G_{mn}^{MM}a_{Mn}\right),$$

where

$$a_{sm}^0 = e^{-ikr_{1s}\cos(\varphi_{1s}-\varphi_0)}i^m\frac{k}{4}\int_0^{2\pi}\left[\rho_s(\varphi_s)\cos(\varphi_s - \varphi_0) + \rho_s'(\varphi_s)\sin(\varphi_{st} - \varphi_0)\right],$$

$$\times J_m(k\rho_s)e^{-ik\rho_s(\varphi_s)\cos(\varphi_s-\varphi_0)-im\varphi_s}d\varphi_s$$

$$(5.53)$$

$$G_{mn}^{ss} = \frac{i}{4} \int_0^{2\pi} J_m(k\rho_s) \left[k\rho_s(\varphi) H_n^{(2)\prime}(k\rho_s) - in \frac{\rho_s'(\varphi)}{\rho_s(\varphi)} H_m^{(2)}(k\rho_s) \right] e^{i(n-m)(\varphi-\pi/2)} d\varphi$$

$$+ \sum_{j=-\infty}^{\infty} (1 - \delta_{j0}) e^{-ijkb_s \cos\varphi_0} \frac{i^{m-n+1}}{4} \sum_{p=-\infty}^{\infty} H_{n-p}^{(2)}(|j|kb_s) e^{i(n-p)\varphi_{j0}}$$

$$\times \int_0^{2\pi} J_m(k\rho_s) \left[k\rho_s(\varphi) J_p'(k\rho_s) - ip \frac{\rho_s'(\varphi)}{\rho_s(\varphi)} J_p(k\rho_s) \right] e^{i(p-m)\varphi} d\varphi, \quad s = 1,\dots,M$$

$$\tag{5.54}$$

$$G_{mn}^{st} = \sum_{j=-\infty}^{\infty} e^{-ijkb_t \cos\varphi_0} \frac{i^{m-n+1}}{4} \sum_{p=-\infty}^{\infty} H_{n-p}^{(2)}\left(kb_{ts}^j\right) e^{i(n-p)\varphi_{ts}^j}$$

$$\times \int_0^{2\pi} J_m(k\rho_s) \left[k\rho_s(\varphi) J_p'(k\rho_s) - ip \frac{\rho_s'(\varphi)}{\rho_s(\varphi)} J_p(k\rho_s) \right] e^{i(p-m)\varphi} d\varphi, \quad s,t = 1,\dots,M; s \neq t,$$

$$\tag{5.55}$$

In these formulas, we have

$$b_{ts}^j = \sqrt{d_{ts}^2 + (jb_t - \delta_{ts})^2}, \quad \cos\varphi_{ts}^j = \frac{\delta_{ts} - jb_t}{b_{ts}^j}, \quad d_{ts} = d_{st}, \quad \delta_{ts} = -\delta_{st}$$

d_{ts} is the distance between tth and sth rows of grating (see Fig. 5.8).

As above, we study the asymptotics of matrix elements and constant terms in system (5.52). For simplicity, we consider only a one-row grating [44]. Proceeding as in Sections 5.1 and 5.2, we can easily show that, for $|m| \gg 1$,

$$|a_m^0| \leq \text{const} \frac{\sigma_1^{|m|}}{|m|! |m|^{1/2}}.$$

Here σ_1 is determined by relations (1.101) and (1.102) for $\alpha = 0$, where φ_0 are the roots of Equations (1.99). Now let us estimate the matrix elements $G_{mn}^{ss} \equiv G_{mn}$. For this, we rewrite relation (5.54) as [44]

$$G_{mn} = G_{mn}^{00} + \sum_{j=-\infty}^{\infty} (1 - \delta_{0j}) \exp(-ijkb \cos\varphi_0) G_{mn}^{0j} \equiv G_{mn}^{00} + G_{mn}^{0\Sigma}.$$

The notation is obvious. For $|m| \gg 1$, we have (see (5.41))

$$|G_{mn}^{00}| \leq \text{const} \frac{\sigma_1^{|m|}}{|m|! |m|^{1/2}}.$$

Similarly, because (see (5.54))

$$G_{mn}^{0\Sigma} = \frac{i}{4} \int_0^{2\pi} J_m(k\rho) e^{i(n-m)(\varphi-\pi/2)} f_n(\varphi) d\varphi,$$

where

$$f_n(\varphi) = \sum_{j=-\infty}^{\infty} \left(1 - \delta_{j0}\right) e^{-ijkb\cos\varphi_0} D\left[H_n^{(2)}\left(kr_j\right) e^{in\left(\varphi_j - \pi/2\right)}\right],$$

and D is the differentiation operator (see above), we estimate the integral asymptotically for $|m| \gg |n|$ and obtain

$$\left|G_{mn}^{0\Sigma}\right| \leq \mathrm{const}\frac{\sigma_1^{|m|}}{|m|!|m|^{1/2}}.$$

So finally, for $|m| \gg |n|$, we have

$$\left|G_{mn}\right| \leq \mathrm{const}\frac{\sigma_1^{|m|}}{|m|!|m|^{1/2}}.$$

Now let $|n| \gg |m|$. Obviously,

$$\left|G_{mn}\right| \leq + \left|G_{mn}^{00}\right| + \left|G_{mn}^{0\Sigma}\right|.$$

Further, for $|n| \gg |m|$, we have (see (5.18))

$$\left|G_{mn}^{00}\right| \leq \mathrm{const}\frac{|n|!}{\sigma_2^{|n|}|n|^{3/2}},$$

where σ_2 is determined by (5.19). For

$$G_{mn}^{0\Sigma} = \sum_{j=-\infty}^{\infty} \left(1 - \delta_{0j}\right) \exp(-ijkb\cos\varphi_0) G_{mn}^{0j},$$

we have

$$\left|G_{mn}^{0\Sigma}\right| \leq \sum_{j=-\infty}^{\infty} \left(1 - \delta_{0j}\right)\left|G_{mn}^{0j}\right|,$$

and by analogy with Section 5.2, we obtain

$$\left|G_{mn}^{0j}\right| \leq \mathrm{const}\frac{|n|!}{\sigma_{0j}^{|n|}|n|^{3/2}},$$

where

$$\sigma_{0j} = \frac{1}{2}\min_{\varphi_0, s}\left|jkbe^{-is\varphi_{j0}} + k\rho(\varphi_0)e^{-is\varphi_0}\right|,$$

i.e., $2\sigma_{0j}/k$ is the distance from the origin O_j in the jth element to the nearest singularity of the central (isolated) scatterer, i.e., to the nearest-to-O_j point in the set of points which are determined by Equations (1.99) and lie inside S. From the last three formulas, we derive

$$|G^{0\Sigma}_{mn}| \leq \text{const}\frac{|n|!}{|n|^{3/2}}\left[\frac{1}{\sigma^{|n|}_{01}}\sum_{j=1}^{\infty}\left(\frac{\sigma_{01}}{\sigma_{0j}}\right)^{|n|} + \frac{1}{\sigma^{|n|}_{0,-1}}\sum_{j=1}^{\infty}\left(\frac{\sigma_{0,-1}}{\sigma_{0,-j}}\right)^{|n|}\right].$$

Each of the sums on the right-hand side of this relation does not exceed $1 + O\left(\frac{1}{|n|}\right)$. For example,

$$\sum_{j=1}^{\infty}\left(\frac{\sigma_{01}}{\sigma_{0j}}\right)^{|n|} = \zeta(|n|)\sum_{j=2}^{\infty}\left[\left(\frac{\sigma_{01}}{\sigma_{0j}}\right)^{|n|} - \frac{1}{j^{|n|}}\right],$$

where $\zeta(|n|)$ is the Riemann zeta function. For $|n|\gg 1$, $\zeta(|n|) < 1 + \frac{1}{|n|}$ [140]. It follows from the formula for σ_{0j} that the sum on the right-hand side of the last relation is of order $O(1/|n|^2)$. Thus, for $|n|\gg|m|$, we have

$$|G^{0\Sigma}_{mn}| \leq \text{const}\frac{|n|!}{|n|^{3/2}}\frac{1}{\sigma^{|n|}_{\Sigma}}, \quad \sigma_{\Sigma} = \min\left(\sigma_{01}, \sigma_{0,-1}\right),$$

and with the estimate for $|G^{00}_{mn}|$ taken into account, we finally obtain

$$|G^{0\Sigma}_{mn}| \leq \text{const}\frac{|n|!}{|n|^{3/2}}\frac{1}{\sigma^{|n|}_{g}}, \quad \sigma_g = \min\left(\sigma_2, \sigma_{\Sigma}\right).$$

In the other situations, the estimate of the matrix elements G_{mn} is similar to that obtained in Section 5.2.

Now we consider a special case of multirow grating of circular cylinders. For $\rho_s(\varphi_s) = a_s = \text{const}$, $s = 1, \ldots, M$, it follows from (5.53)–(5.55) that

$$a^0_{sm} = \frac{i\pi ka_s}{2}e^{-ikr_{1s}\cos(\varphi_{1s}-\varphi_0)}J_m(ka_s)J'_m(ka_s)e^{-im\varphi_0}, \quad s = 1, \ldots, M,$$

$$G^{ss}_{mn} = \frac{i\pi ka_s}{2}J_m(ka_s)\Big[H^{(2)\prime}_m(ka_s)\delta_{mn}$$

$$+ J'_m(ka_s)\sum_{j=-\infty}^{\infty}(1-\delta_{j0})e^{-ijkb_s\cos\varphi_0}H^{(2)}_{n-m}(|j|kb_s)e^{i(n-m)(\varphi_{j0}-\pi/2)}\Big], \quad s = 1, \ldots, M,$$

$$G^{st}_{mn} = \frac{i\pi ka_s}{2}J_m(ka_s)J'_m(ka_s)\sum_{j=-\infty}^{\infty}e^{-ijkb_t\cos\varphi_0}H^{(2)}_{n-m}\left(kb^j_{ts}\right)e^{i(n-m)(\varphi^j_{ts}-\pi/2)},$$

$s, t = 1, \ldots, M$, $s \neq t$.

$$(5.56)$$

For the grating of circular cylinders in the one-mode approximation (for $ka_s \ll 1$, $s = 1, \dots, M$), system (5.52) becomes

$$\left(1 - G_{00}^{11}\right) a_{10} - G_{00}^{12} a_{20} - \cdots - G_{00}^{1M} a_{M0} = a_{10}^0,$$
$$- G_{00}^{21} a_{10} + \left(1 - G_{00}^{22}\right) a_{20} - \cdots - G_{00}^{2M} a_{M0} = a_{20}^0,$$
$$- G_{00}^{M1} a_{10} - G_{00}^{M2} a_{20} - \cdots + \left(1 - G_{00}^{MM}\right) a_{M0} = a_{M0}^0,$$

where we now have (see (5.56))

$$G_{00}^{ss} = \frac{i\pi k a_s}{2} J_0(k a_s) \left[H_0^{(2)\prime}(k a_s) + J_0'(k a_s) \sum_{j=-\infty}^{\infty} \left(1 - \delta_{j0}\right) e^{-ijk b_s \cos\varphi_0} H_{n-m}^{(2)}(|j|k b_s) \right],$$

$$G_{00}^{st} = \frac{i\pi k a_s}{2} J_0(k a_s) J_0'(k a_s) \sum_{j=-\infty}^{\infty} e^{-ijk b_t \cos\varphi_0} H_0^{(2)}\left(k b_{ts}^j\right), \quad s, t = 1, \dots, M, s \neq t.$$

Figure 5.9 shows how the reflection and transmission coefficients depend on the frequency in the case of a 10-row grating of equal ideally reflecting circular cylinders [141]. The following geometric parameters were used: $a/b = 0.01$, $h/b = 0.03$ (where h is the distance between the rows) and the angle of the wave incidence $\theta_0 = 45°$. The dashed curve in the figure corresponds to the asymptotic solution of the problem obtained by formulas (5.51), and the solid curve corresponds to the rigorous solution obtained by relations (5.52)–(5.56). Figure 5.10 shows the dependence of the plane wave reflection and transmission coefficients on the angle of incidence of the plane wave

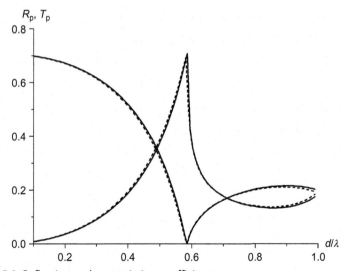

Figure 5.9 Reflection and transmission coefficients.

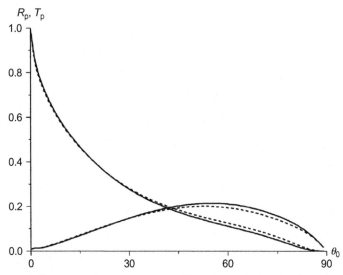

Figure 5.10 Reflection and transmission coefficients on the angle of incidence of the plane wave.

for the same parameters of the model. The curves are denotes as in the preceding figure. The curves are constructed for the wave parameter $b/\lambda = 0.99$.

5.4 SOLUTION OF THE THREE-DIMENSIONAL ACOUSTIC PROBLEM OF DIFFRACTION AT A COMPACT SCATTERER

So we consider the boundary-value problem

$$\Delta U^1 + k^2 U^1 = 0, \tag{5.57}$$

$$U|_S = \frac{W}{k} \frac{\partial U}{\partial n}\bigg|_S, \tag{5.58}$$

where U^1 is the potential of the scattered acoustic field velocity, $U = U^0 + U^1$ is the potential of the total field velocity, U^0 is the potential of the initial or incident field velocity, $k = \frac{\omega}{c}$, $W = \frac{Z}{i c \rho}$, c is the speed of sound, ρ is the medium density, Z is the local acoustic impedance which is constant on S, S is the scatterer boundary, and $\frac{\partial}{\partial n}$ denotes the differentiation in the direction of the outer normal on S.

In addition to (5.57) and (5.58), the function U^1 must also satisfy the radiation condition (see above).

In what follows, for brevity, the function U^1 is called the wave field or diffraction field as previously.

In the wave region (i.e., in the domain $r \gg d$, where d is the greatest dimension of the body), we have the asymptotic relation

$$U^1(r, \theta, \varphi) = \frac{e^{-ikr}}{kr} g(\theta, \varphi) + O\left(\frac{1}{(kr)^2}\right), \tag{5.59}$$

where $g(\theta, \varphi)$ is the scattering pattern. With the boundary condition (5.58) taken into account, we can obtain the following representation for the function $g(\theta, \varphi)$ in spherical coordinates:

$$g(\alpha, \beta) = \int_0^{2\pi} \int_0^{\pi} v(\theta, \phi) \left\{ 1 - i\frac{W}{\chi} \left[R(\theta, \phi)\cos\gamma \sin\theta - R'_\theta(\theta, \phi)\frac{\partial \cos\gamma}{\partial \theta}\sin\theta \right.\right.$$
$$\left.\left. - R'_\phi(\theta, \phi)\frac{\partial \cos\gamma}{\partial \phi}\frac{1}{\sin\theta} \right] \right\} \exp[ikR(\theta, \phi)\cos\gamma] d\theta d\phi \tag{5.60}$$

where

$$v(\theta, \phi) = -\frac{k}{4\pi}\chi R(\theta, \phi)\frac{\partial u}{\partial n} = -\frac{k}{4\pi}\left[R^2(\theta, \phi)\sin\theta\frac{\partial u}{\partial r} \right.$$
$$\left. - R'_\theta\frac{\partial u}{\partial \theta}\sin\theta - \frac{R'_\phi}{\sin\theta}\frac{\partial u}{\partial \phi} \right]\Bigg|_{r=R(\theta, \phi)}, \tag{5.61}$$

$$\chi = \sqrt{[R^2(\theta, \phi) + R'^2_\theta(\theta, \phi)]\sin^2\theta + R'^2_\phi(\theta, \phi)}, \quad dS = \chi R d\theta d\phi,$$

$$\cos\gamma = [\sin\theta\sin\alpha\cos(\phi - \beta) + \cos\theta\cos\alpha],$$

and $r = R(\theta, \phi)$ is the equation of the surface S.

For the wave field, we use representation (1.38),

$$U^1(r, \theta, \varphi) = \frac{1}{2\pi i}\int_0^{2\pi} \int_0^{\pi/2 + i\infty} \exp(-ikr\cos\alpha)\hat{g}(\alpha, \beta; \theta, \varphi)\sin\alpha \, d\alpha d\beta, \tag{5.62}$$

where the generalized pattern of the wave field $\hat{g}(\alpha, \beta; \theta, \varphi)$ in the case under study is determined by the relation

$$\hat{g}(\alpha, \beta; \theta, \varphi) = \int_0^{2\pi} \int_0^{\pi} v(\theta', \varphi') \left\{ 1 - i\frac{W}{\chi}\left[R(\theta', \varphi')\cos\hat{\gamma}\sin\theta' - R'_{\theta'}(\theta', \varphi')\frac{\partial\cos\hat{\gamma}}{\partial\theta'}\sin\theta' \right.\right.$$
$$\left.\left. - R'_{\varphi'}(\theta', \varphi')\frac{\partial\cos\hat{\gamma}}{\partial\varphi'}\frac{1}{\sin\theta'} \right] \right\}\exp[ikR(\theta', \varphi')\cos\hat{\gamma}] d\theta' d\varphi' \tag{5.63}$$

and

$$\cos\hat{\gamma} = \{\sin\theta'\sin(\varphi'-\varphi)\sin\alpha\cos\beta + [\sin\theta\cos\theta' - \cos\theta\sin\theta'\cos(\varphi'-\varphi)]$$

$$\times \sin\alpha\sin\beta + [\cos\theta\cos\theta' + \sin\theta\sin\theta'\cos(\varphi'-\varphi)]\cos\alpha\} = \vec{p}^{\mathrm{T}}B\vec{i}_{r'},$$

where (see Chapter 1)

$$\vec{p}^{\mathrm{T}} = (\sin\alpha\cos\beta, \sin\alpha\sin\beta, \cos\alpha), \quad \vec{i}_{r'} = \vec{r'}\big/\left|\vec{r'}\right|,$$

$$B = \begin{pmatrix} -\sin\varphi & \cos\varphi & 0 \\ -\cos\varphi\cos\theta & -\sin\varphi\cos\theta & \sin\theta \\ \cos\varphi\sin\theta & \sin\varphi\sin\theta & \cos\theta \end{pmatrix}$$

is the matrix of the coordinate system rotation such that the axis z coincides with the direction to the observation point.

Representation (5.62) allows us to determine the function U^1 everywhere in $\mathbb{R}^3/\bar{B}_0$, where $\bar{B}_0$ is the convex envelope of the singularities of the wave field $U^1(r,\theta,\varphi)$. As above, we use relations (5.60)–(5.62) to obtain the integral-operator equation of the second type for the scattering pattern

$$g(\alpha,\beta) = g^0(\alpha,\beta) + \frac{1}{8\pi^2}\int_0^{2\pi}\int_0^{\pi}\int_0^{2\pi}\int_0^{\pi/2+i\infty}\left\{1 - i\frac{W}{\chi}[R(\theta,\varphi)\right.$$

$$\times\cos\gamma\sin\theta - R_\theta'\frac{\partial\cos\gamma}{\partial\theta}\sin\theta - \frac{R_\varphi'}{\sin\theta}\frac{\partial\cos\gamma}{\partial\varphi}\bigg]\bigg\}\bigg[ik^2R^2(\theta,\varphi)\cos\alpha'$$

$$\times\hat{g}(\alpha',\beta';\theta,\varphi)\sin\theta + \hat{g}_\theta{}'(\alpha',\beta';\theta,\varphi)kR_\theta'\sin\theta + \hat{g}_\varphi{}'(\alpha',\beta';\theta,\varphi)\frac{kR_\varphi'}{\sin\theta}\bigg]$$

$$\times\exp\left[ikR(\theta,\varphi)(\cos\gamma - \cos\alpha')\right]\sin\alpha'\,d\alpha'\,d\beta'\,d\theta d\varphi. \tag{5.64}$$

In this equation, $g^0(\alpha,\beta)$ is a known function determined by (5.60), where the function $v(\theta,\varphi)$ must be replaced by

$$v^0(\theta,\varphi) = -\frac{k}{4\pi}\chi R(\theta,\varphi)\frac{\partial U^0}{\partial n}.$$

One can show (see [95]) that Equation (5.64) is solvable under the condition that

$$\bar{B}_0 \subseteq \bar{D}, \tag{5.65}$$

where $\bar{D}$ is the domain occupied by the scatterer together with its boundary.

In the special case where the scatter is a ball (i.e., $R(\theta, \varphi) = a \equiv \text{const}$), Equation (5.64) becomes

$$g(\alpha, \beta) = g^0(\alpha, \beta) + \frac{(ka)^2}{8\pi^2} \int_0^{2\pi} \int_0^{\pi} \int_0^{2\pi} \int_0^{\pi/2+i\infty} (1 - iW\cos\gamma)\cos\alpha'$$

$$\times \exp[ika(\cos\gamma - \cos\alpha')]\hat{g}(\alpha', \beta'; \theta, \varphi)\sin\alpha'\, d\alpha'\, d\beta'\sin\theta d\theta d\varphi.$$

$$(5.66)$$

Equations (5.64) and (5.66) are integral-operator equations, i.e., their integrand contains not the desired function $g(\theta, \varphi)$ but the result of its transformation by an operator related to the above-mentioned rotation of the coordinate system (cf. formulas (5.60) and (5.63)). Nevertheless, Equations (5.64) and (5.66) also preserve many properties of the Fredholm integral equations. In particular, for small values of the parameter kd, these equations can be solved by the iteration method. For example, solving Equation (5.66) by this method, we obtain

$$g(\alpha, \beta) = -i\sum_{n=0}^{\infty}(2n+1)\frac{j_n(ka) - Wj_n'(ka)}{h_n^{(2)}(ka) - Wh_n^{(2)\prime}(ka)}P_n(\cos\alpha). \qquad (5.67)$$

Expression (5.67) coincides with the well-known solution for the impedance sphere. The series (5.67) converges absolutely for all values of the parameter ka, and hence it can also be used to construct asymptotic (for $ka \gg 1$) solutions [95].

In the general case of nonspherical scatterers, Equation (5.64) can also be solved by the iteration method, but in this case, the solution cannot be written in the closed analytic form. Therefore, as above, we solve Equation (5.64) by a different method, i.e., we reduce it to a system of algebraic equations. For this, we use the expansion of the pattern in the generalized Fourier series:

$$g(\theta, \varphi) = \sum_{n=0}^{\infty}\sum_{m=-n}^{n} a_{nm}P_n^m(\cos\theta)e^{im\varphi} \qquad (5.68)$$

It follows from (5.63) that

$$\hat{g}(\alpha, \beta; \theta, \varphi) = \sum_{n=0}^{\infty}\sum_{m=-n}^{n} \hat{a}_{nm}(\theta, \varphi)P_n^m(\cos\alpha)e^{im\beta}, \qquad (5.69)$$

where

$$\hat{a}_{nm}(\theta, \varphi) = \sqrt{\frac{(n-m)!}{(n+m)!}}\sum_{s=-n}^{n}\sqrt{\frac{(n+s)!}{(n-s)!}}a_{ns}i^s P_{ms}^n(\cos\theta)e^{is\varphi},$$

and $P_{ms}^n(\cos\theta)$ are generalized Legendre functions [39].

Substituting (5.68) and (5.69) into (5.64) and using the relation [39]

$$P_{0s}^n(z) = (-i)^s \sqrt{\frac{(n-s)!}{(n+s)!}} P_n^s(z),$$

we obtain the system of algebraic equations for the coefficients a_{nm}:

$$a_{nm} = a_{nm}^0 + \sum_{\nu=0}^{\infty} \sum_{\mu=-\nu}^{\nu} G_{nm,\nu\mu} a_{\nu\mu}, \quad |m| \le n, \quad n = 0, 1, \ldots \quad (5.70)$$

where

$$
\begin{aligned}
G_{nm,\nu\mu} = {}& i^{n-\nu}(2n+1)\frac{(n-m)!}{(n+m)!}\frac{i}{4\pi}\int_0^{2\pi}\int_0^\pi \Big[k^2 R^2(\theta,\varphi) h_\nu^{(2)\prime}(kR) P_\nu^\mu(\cos\theta)\sin\theta \\
& -h_\nu^{(2)}(kR)(\sin\theta)kR_\theta' P_\nu^\mu(\cos\theta) - i\mu\frac{kR_\varphi'}{\sin\theta}h_\nu^{(2)}(kR)P_\nu^\mu(\cos\theta) \Big] \\
& \times\Big\{ j_n(kR)P_n^m(\cos\theta) - \frac{iW}{\chi(\theta,\varphi)}\Big[-iR(\theta,\varphi)(\sin\theta)j_n'(kR)P_n^m(\cos\theta) \\
& + \frac{iR_\theta'}{kR}j_n(kR)(\sin\theta)\frac{d}{d\theta}P_n^m(\cos\theta) + \frac{mR_\varphi'}{kR\sin\theta}j_n(kR)P_n^m(\cos\theta) \Big] \Big\} e^{i(\mu-m)\varphi}\,d\theta\,d\varphi
\end{aligned}
$$

$$(5.71)$$

The constant terms a_{nm}^0 are coefficients of the expansion of the known function $g^0(\alpha,\beta)$:

$$
\begin{aligned}
a_{nm}^0 = {}& i^n(2n+1)\frac{(n-m)!}{(n+m)!}\int_0^{2\pi}\int_0^\pi v^0(\theta,\varphi)\Big\{ j_n(kR(\theta,\varphi))P_n^m(\cos\theta) \\
& -\frac{W}{\chi kR(\theta,\varphi)}\Big[kR^2(\theta,\varphi)j_n'(kR)P_n^m(\cos\theta)\sin\theta - R_\theta'(\theta,\varphi)j_n(kR)\frac{dP_n^m}{d\theta}\sin\theta \\
& +im\frac{R_\varphi'(\theta,\varphi)}{\sin\theta}j_n(kR)P_n^m(\cos\theta) \Big] \Big\} e^{-im\varphi}\,d\theta\,d\varphi
\end{aligned}
$$

An important advantage of system (5.70) is the fact that, as follows from (5.71), its matrix elements can be expressed in terms of integrals with dimensions about twice smaller than those of the widely used methods such as the method of current integral equation, the method of moments, etc. Moreover, as we see below, the solution of system (5.70) obtained by the reduction method is characterized a rather high rate of convergence weakly depending on the scatterer geometry.

In the case $W = 0$ (acoustically soft body) or in the case $W = \infty$ (acoustically hard body), the initial boundary-value problem becomes the Dirichlet or Neumann problem, respectively. In both cases, the algebraic system has the same form (5.70), and we can show that

$$G^D_{nm,\nu\mu} + G^N_{nm,\nu\mu} = \delta_{\nu n}\delta_{\mu m}, \quad a^{0D}_{nm} + a^{0N}_{nm} = 0.$$

The notation is obvious. The obtained relationship between the elements of algebraic systems of the Dirichlet and Neumann problems is very convenient from the computational standpoint.

If the scatterer is a body of revolution, i.e., if $R(\theta, \phi) = R(\theta)$, then the algebraic system (5.70) becomes

$$a_{nm} = a^0_{nm} + \sum_{\nu=|m|}^{\infty} G_{nm,\nu m} a_{\nu m}, \quad m = 0, \pm 1, \ldots, \pm n; n = 0, 1, \ldots. \quad (5.72)$$

In this system,

$$G_{nm,\nu m} = i^{n-\nu}(2n+1)\frac{(n-m)! \, i}{(n+m)! \, 2}\int_0^\pi \left\{ j_n(kR)P_n^m(\cos\theta) - \frac{W}{\chi(\theta)}\left[R(\theta)j_n'(kR)\right.\right.$$

$$\times P_n^m(\cos\theta) - \frac{R_\theta'}{kR(\theta)}j_n(kR)\frac{\mathrm{d}}{\mathrm{d}\theta}P_n^m(\cos\theta)\bigg]\sin\theta\bigg\}$$

$$\times \left[k^2 R^2(\theta)h_\nu^{(2)'}(kR)P_\nu^m(\cos\theta) - kR_\theta' h_\nu^{(2)}(kR)\frac{\mathrm{d}}{\mathrm{d}\theta}P_\nu^m(\cos\theta)\right](\sin\theta)\mathrm{d}\theta$$

$$(5.73)$$

Proceeding as in Ref. [94], we can show that, for $n \gg \nu$,

$$|G_{nm,\nu m}| \le \mathrm{const}\frac{\sigma_1^{\,n}}{n \cdot n!}, \quad (5.74)$$

where

$$\sigma_1 = \max_{\theta_0,\phi_0,s}\left|\frac{kR(\theta_0,\phi_0)}{2}e^{is\theta_0}\right|, \quad (5.75)$$

and θ_0, ϕ_0 are roots of the equations

$$\left[R_\theta'(\theta,\phi) + isR(\theta,\phi)\right]e^{is\theta} = 0; \quad s = \pm 1; R_\phi'(\theta,\phi) = 0. \quad (5.76)$$

The maximum in (5.75) is sought in the set of roots of system (5.76) which, after the change of variables $\varsigma = R(\theta,\phi)e^{i\theta}$, lie inside the contours C_ϕ. These contours are the mappings of the surface S cross-section by the plane $(\phi, \phi + \pi)$ onto the plane $z = re^{i\alpha}$.

For $\nu \gg n$, we can similarly show that

$$|G_{nm,\nu\mu}| \leq \text{const} \frac{\nu!}{\sigma_2{}^\nu},$$

and

$$\sigma_2 = \min_{\theta_0, \phi_0, s} \left| \frac{kR(\theta_0, \phi_0)}{2} e^{is\theta_0} \right|, \qquad (5.77)$$

while the minimum in this relation is sought in the set of roots of system (5.76), which after the change of variables $\varsigma = Re^{i\theta}$ are associated with the points outside the above-described contour C_ϕ on the plane $z = re^{i\alpha}$.

For $n \gg 1$, we can similarly show that

$$|a^0{}_{nm}| \leq \text{const} \frac{\sigma^n}{n \cdot n!}, \quad \sigma = \max(\sigma_1, \sigma_0), \qquad (5.78)$$

$\sigma_0 = \frac{kr_0}{2}$, and r_0 is the distance to the point inside S which is most remote from the origin and corresponds to the singularity of the function $v^0(\theta, \phi)$ arising when it is analytically continued into the region of complex angles [9] (also see Chapter 1). We note that if U^0 is the plane wave field, then $\sigma_0 = 0$.

It follows from the above estimates that it is necessary to replace the coefficients [94] in system (5.70) by setting

$$a_{nm} = \frac{\sigma^n}{n!} x_{nm}. \qquad (5.79)$$

After this change, system (5.70) becomes

$$x_{nm} = x^0{}_{nm} + \sum_{\nu=0}^{\infty} \sum_{\mu=-\nu}^{\nu} g_{nm,\nu\mu} x_{\nu\mu}, \quad |m| \leq n, n = 0, 1, \ldots, \qquad (5.80)$$

where

$$x^0{}_{nm} = \frac{n!}{\sigma^n} a_{nm}, \quad g_{nm,\nu\mu} = \frac{n!}{\nu!} \sigma^{\nu-n} G_{nm,\nu\mu}.$$

System (5.80) can be solved by the reduction method under the condition that

$$\sigma_2 > \sigma. \qquad (5.81)$$

We note that relations (5.65) and (5.81) determine the restrictions on the parameters of the boundary-value problem under study, and the proposed approach is rigorous under these restrictions, i.e., it principally permits obtaining the solution with a prescribed accuracy.

The wave scattering at bodies that do not satisfy condition (5.81), in particular, at significantly nonconvex bodies, can be modeled by a group of bodies of a simpler geometry whose configuration repeats the geometry of the initial scatterer (see above). The diffraction problem at a strongly nonconvex object by a group or bodies is modeled on the basis of a system of equations of the form (5.24) whose number is equal to the number of bodies. For the solvability of these equations, it is sufficient that each body is weakly nonconvex and that the bodies do not contact with each other. In this case, as previously noted, the rate of convergence of the PEM computational algorithm for a group of bodies is practically independent of the distances between the bodies until they begin to contact.

When solving the wave-scattering problems by such a universal method as the method of integral equations for the field on the body surface [5,8,11,12], the corresponding algebraic systems, even in the two-dimensional case, are of the order of $10 \div 20\,kd$, where d is the characteristics dimension of the body [8]. However, in the three-dimensional case, the system size is obviously proportional to the squared value, and the presence of break points on the boundary obviously makes the algorithm more complicated.

Numerical calculations were used to study the problems of wave scattering at the following axially symmetric bodies: a prolate spheroid with the major semiaxis c (along the axis z) and minor semiaxis a, a circular cylinder of radius a and height $2c$, a cylinder with hemispheres at the ends such that the cylindrical part is of height h and the hemisphere radii are equal to the cylinder radius a, and a circular cone with spherical base such that the base radius a is equal to the cone radius and the height of the conic part is h. The calculations were performed for the following three values of the impedance: $W = 0$, $W = 1000$, $W = -i$. The last of these values corresponds to the case of the so-called consistent impedance [8].

The case of a unit plane wave incident at the angles $\theta_0 = \pi/2$, $\varphi_0 = 0$ was considered for the above-listed bodies except the cone for which the wave was incident along the cone axis, i.e., at the angles $\theta_0 = \varphi_0 = 0$. The scattering pattern was calculated depending on the angle θ in the plane $\varphi = 0, \pi$.

In all calculations, we solved a finite system of the form

$$x_{nm} = x_{nm}^0 + \sum_{\nu=|\mu|}^{N} g_{nm,\nu m} x_{\nu m}, \ |m| \le n, \ n = 0, 1, \dots, N,$$

(which follows from the general system (5.80) in the case of a body of revolution), where N was equal to 15, i.e., $N \approx 2kd$, in all calculations (except for Table 5.2).

Figures 5.11–5.13 show the scattering patterns for a circular cylinder of height $kh = 4$ (curves 1) and the "inscribed" prolate spheroid with semiaxes $ka = 2$, $kc = 4$ (curves 2). In this case, Fig. 5.11 corresponds to the impedance $W = 1000$, Fig. 5.12, to $W = 0$, and Fig. 5.13, to $W = -i$. One can see that the greatest difference between the patterns is observed in the case of acoustically hard scatterer. In the scattering patterns, the main lobe is the shadow lobe whose width agrees well with the estimates of geometrical optics. However, in the pattern of an acoustically hard scatterer, the back lobe level is already sufficiently close to the shadow lobe level. In the case of a scatterer

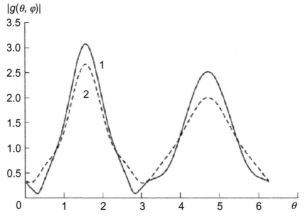

Figure 5.11 Scattering patterns for a circular cylinder of height $kh = 4$ and the "inscribed" prolate spheroid with semiaxes $ka = 2$, $kc = 4$, $W = 1000$.

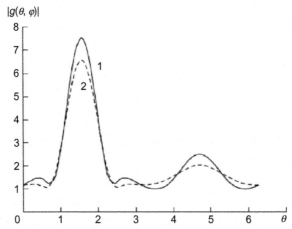

Figure 5.12 Scattering patterns for a circular cylinder of height $kh = 4$ and the "inscribed" prolate spheroid with semiaxes $ka = 2$, $kc = 4$, $W = 0$.

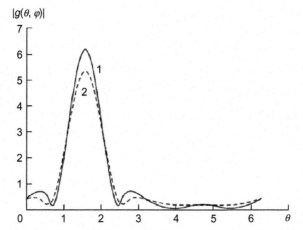

Figure 5.13 Scattering patterns for a circular cylinder of height $kh=4$ and the "inscribed" prolate spheroid with semiaxes $ka=2$, $kc=4$, $W=-i$.

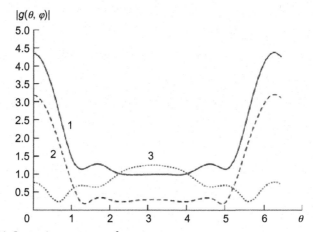

Figure 5.14 Scattering patterns of a cone.

with consistent impedance, the shadow lobe is significantly (approximately by 30 dB) greater than that of the back lobe, while in the case of an acoustically soft scatterer, it is greater only approximately by 10 dB. This picture agrees well with the intuitive concepts of "ideally absorbing" scatterer.

Figure 5.14 shows the scattering patterns of a cone characterized by the geometric parameters $ka=2$ and $kh=6$ for different values of the surface impedance: curve 1 corresponds to $W=0$, curve 2, to $W=-i$, and curve 3, to $W=1000$. Here the back lobe of the pattern of the cone with consistent impedance is approximately by 20 dB less than the shadow lobe. One can see that the scattering pattern is close to constant on a rather large

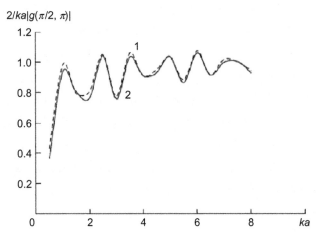

Figure 5.15 Parameter $\frac{2}{ka}|g(\pi/2,\pi)|$.

segment (of the order of 90°), which corresponds to the scattering at the spherical part of the body.

Figure 5.15 illustrates the results of calculations of the parameter $\frac{2}{ka}|g(\pi/2,\pi)|$, i.e., of the relative radius of the so-called equivalent sphere [8] characterizing the inverse scattering cross-section of the body depending on ka for the sphere (curve 2) and the cylinder with spherical bases and of height $kh = 0.1$, i.e., for a cylinder slightly different from the sphere, for the impedance $W = 1000$. One can see that the difference between these two curves is small (less than 2%).

Figure 5.16 shows the results of similar calculations for $\frac{2}{kc}|g(\pi/2,\pi)|$ as a function of kc for the cylinder (curve 1) and spheroid (curve 2) with the same geometric parameters as in Figs. 5.11–5.13 for $W = 1000$. Here the parameter c is equal to half the length of the body. One can see that both curves vary synchronously, but the curve for the cylinder is here noticeably higher, which corresponds to Fig. 5.11 and can be explained by the fact that a significant part of the cylinder surface has a rectilinear generator normal to the incident wave.

Table 5.1 presents the data illustrating the optical theorem accuracy in the statement given in Ref. [8] (see formula (3.23)).

All calculations were performed for bodies whose scattering patterns are shown in Figs. 5.11–5.13 for $W = 0$.

Finally, Table 5.2 contains the data characterizing the rate of convergence of the computational algorithm for the cylinder, spheroid, and cone. One can see that in the case of a body with an analytic boundary (spheroid),

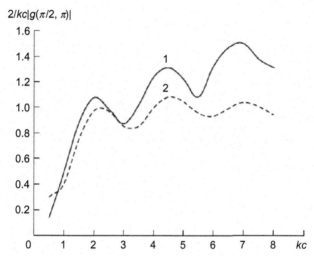

Figure 5.16 Parameter $\frac{2}{kc}|g(\pi/2,\pi)|$.

Table 5.1 Optical theorem accuracy.

	Spheroid	Cylinder	Cone
$k^2\sigma_S/4\pi$	3,78528	2,94096	2,97451
$-\mathrm{Im}g(0,0)$	3,78528	2,94357	2,97453

the two true decimal digits are established already for $N \approx kd$. Approximately the same rate of convergence is obtained when we solve the problem of wave diffraction at a spheroid by the method of separation of variables using the expansions in spheroidal functions [142]. For bodies with boundary breaks (cylinder, cone), it is necessary to take $N \approx 2kd$ in calculations, which is also several times (and sometimes several orders of magnitude) better than the similar characteristics of the traditional methods.

The above-listed numerical results allow us to conclude that the PEM is a rather efficient tool for solving the problems of wave diffraction at impedance bodies of complex geometry including the bodies with edges and conic points. The computational algorithm constructed on the basis of PEM is a highly fast operating algorithm. In this case, the calculations of the scattering characteristics of smooth bodies and bodies with singularities (edges, etc.) are performed in the framework of the same algorithm without any special measures.

Table 5.2 Rate of convergence of the computational algorithm.

	N	9	11	13	15	17
Cylinder (ka = 2, kh = 4)						
$W = 0$	$g(\pi/2,0)$	7.5025×10	7.5172×10	7.5229×10	7.5200×10	7.4486×10
$W = 0$	$g(\pi/2,\pi)$	2.5663×10	2.5169×10	2.4979×10	2.5026×10	2.5027×10
$W = 1000$	$g(\pi/2,0)$	3.0442×10	3.0743×10	3.0838×10	3.0653×10	3.0644×10
$W = 1000$	$g(\pi/2,\pi)$	2.4335×10	2.4788×10	2.5108×10	2.5070×10	2.5066×10
Spheroid (ka = 2, kc = 4)						
$W = 0$	$g(\pi/2,0)$	6.5647×10	6.5649×10	6.5649×10	6.5659×10	6.5653×10
$W = 0$	$g(\pi/2,\pi)$	2.0405×10	2.0404×10	2.0404×10	2.0413×10	2.0416×10
$W = 1000$	$g(\pi/2,0)$	2.6626×10	2.6615×10	2.6615×10	2.6615×10	2.6615×10
$W = 1000$	$g(\pi/2,\pi)$	1.9921×10	1.9919×10	1.9919×10	1.9919×10	1.9918×10
Spheroid (ka = 1,005; kc = 5,1)						
$W = 0$	$g(\pi/2,0)$	5.7223×10^{-1}	5.7231×10^{-1}	5.7005×10^{-1}	5.7006×10^{-1}	5.7005×10^{-1}
$W = 0$	$g(\pi/2,\pi)$	5.7178×10^{-1}	5.7186×10^{-1}	5.6960×10^{-1}	5.6962×10^{-1}	5.6961×10^{-1}
$W = 1000$	$g(\pi/2,0)$	1.2399×10^{-1}	1.2408×10^{-1}	1.2368×10^{-1}	1.2369×10^{-1}	1.2367×10^{-1}
$W = 1000$	$g(\pi/2,\pi)$	1.2539×10^{-1}	1.2564×10^{-1}	1.2524×10^{-1}	1.2523×10^{-1}	1.2525×10^{-1}
Cone (ka = 2; kh = 6)						
$W = 0$	$g(\pi/2,0)$	6.3459×10^{-1}	6.1876×10^{-1}	6.0257×10^{-1}	6.0523×10^{-1}	6.0350×10^{-1}
$W = 0$	$g(\pi/2,\pi)$	6.5045×10^{-1}	6.9022×10^{-1}	6.7486×10^{-1}	6.7215×10^{-1}	6.7252×10^{-1}
$W = 1000$	$g(\pi/2,0)$	9.6439×10^{-1}	9.4209×10^{-1}	9.2592×10^{-1}	9.2410×10^{-1}	9.2467×10^{-1}
$W = 1000$	$g(\pi/2,\pi)$	7.4507×10^{-1}	7.5850×10^{-1}	7.6272×10^{-1}	7.6862×10^{-1}	7.6735×10^{-1}

5.5 PLANE WAVE SCATTERING AT A PERIODIC INTERFACE BETWEEN MEDIA

The problem of reflection and refraction of a plane electromagnetic wave incident at an angle on a periodic interface between two half-spaces with different material parameters has been studied by several authors. This problem and its different modifications are of great importance in the theory and technology of fiber-optical communication systems (e.g., see [143]) and in other fields of physics and technology. Here we briefly describe the PEM used to solve the problem of plane wave scattering at a "transparent" periodic interface between two media [96]. We consider only the case of normal polarization where the field has the components E_z, H_x, H_y and the wave vector of the incident wave lies in the plane XOY. The case of parallel polarization can be considered similarly.

We assume that the interface is described by the equation $y = h(x)$ and introduce the notation

$$E_z = U_1 \equiv U^0 + U^1, \quad y \geq h(x),$$
$$E_z = U_2, \quad y \leq h(x),$$

where

$$U^0 = \exp(-ik_1 x \sin\theta_1 + ik_1 y \cos\theta_1), \tag{5.82}$$

U^0 is the incident wave and U^1 is the reflected wave in the first media, U_2 is the field in the second media, $k_j = \omega\sqrt{\varepsilon_j \mu_j}$, ε_j, μ_j are the wave number and the absolute dielectric permittivity and magnetic permeability of the jth medium, respectively, $j = 1, 2$; $k = \omega\sqrt{\varepsilon_0 \mu_0}$.

We have (see Section 1.1.3)

$$U^1(x, y) = \frac{2}{b} \sum_{n=-\infty}^{\infty} f_1(w_{1n}) \frac{\exp(-iw_{1n}x - iv_{1n}y)}{v_{1n}}, \quad y > \frac{2\sigma_S^+}{k} \equiv \frac{2\sigma_1}{k}, \tag{5.83}$$

$$U_2(x, y) = \frac{2}{b} \sum_{n=-\infty}^{\infty} f_2(w_{2n}) \frac{\exp(-iw_{2n}x + iv_{2n}y)}{v_{2n}}, \quad y < \frac{2\sigma_S^-}{k} \equiv \frac{2\sigma_2}{k}. \tag{5.84}$$

In these relations, b is the period of the function $h(x)$,

$$w_{jn} = \frac{2\pi}{b} n + k_j \sin\theta_j, \quad j = 1, 2, \tag{5.85}$$

and according to the Floquet theorem (see [45]),

$$k_1 \sin\theta_1 = k_2 \sin\theta_2. \tag{5.86}$$

Further,

$$v_{jn} = \sqrt{k_j^2 - w_{jn}^2}, \quad \mathrm{Im}\, v_{jn} \leq 0, \tag{5.87}$$

$$f_j(w_{jn}) = \frac{1}{4} \int_{-b/2}^{b/2} \left\{ U_j(x,y) \left[v_{jn} \mp w_{jn} h'(x) \right] \pm \left[\frac{\partial U_j}{\partial y} - \frac{\partial U_j}{\partial x} h'(x) \right] \right\} \bigg|_{y=h(x)}$$
$$\times \exp\left[i w_{jn} x \pm i v_{jn} h(x) \right] dx \tag{5.88}$$

the superscripts in these formulas correspond to $j=1$, and the subscripts, to $j=2$. The quantities $\sigma_S^{\pm}$ are determined by the geometric properties of the surface $y = h(x)$ (see Section 1.2.1; also see [45]) and by the relations $\sigma_S^+ = \sigma_S(0)$, $\sigma_S^- = \sigma_S(\pi)$ (see (1.52a)).

First, we assume that $\sigma_1 < k[h(x)]_{\min}/2$ and $\sigma_2 > k[h(x)]_{\max}/2$. Then using relations (5.83), (5.84) and (5.88), we can obtain an algebraic system for the coefficients $f_{jn} \equiv f_j(w_{jn})$. For this, according to the continuity conditions for the tangent components of the electric and magnetic vectors on the interface,

$$[U_1(x,y) - U_2(x,y)]\big|_{y=h(x)} = 0,$$
$$\left\{ \left[\frac{\partial U_1}{\partial y} - \frac{\partial U_1}{\partial x} h'(x) \right] - \frac{\mu_1}{\mu_2} \left[\frac{\partial U_2}{\partial y} - \frac{\partial U_2}{\partial x} h'(x) \right] \right\} \bigg|_{y=h(x)} = 0, \tag{5.89}$$

we express f_{1n} in terms of f_{2n} and conversely

$$f_{1n} = \sum_{m=-\infty}^{\infty} f_{2m} \left\{ \left(\frac{v_{1n}}{v_{2m}} - \frac{\mu_1}{\mu_2} \right) \frac{1}{2b} \int_{-b/2}^{b/2} \exp[ix(w_{1n} - w_{2m}) + i(v_{1n} + v_{2m})h(x)] dx \right.$$
$$\left. - \left(\frac{w_{1n}}{v_{2m}} + \frac{\mu_1}{\mu_2} \frac{w_{2m}}{v_{2m}} \right) \frac{1}{2b} \int_{-b/2}^{b/2} \exp[ix(w_{1n} - w_{2m}) + i(v_{1n} + v_{2m})h(x)]h'(x) dx \right\}, \tag{5.90}$$

$$f_{2n} = f_{0n} + \sum_{m=-\infty}^{\infty} f_{1m} \left\{ \left(\frac{v_{2n}}{v_{1m}} - \frac{\mu_2}{\mu_1} \right) \frac{1}{2b} \int_{-b/2}^{b/2} \exp[ix(w_{2n} - w_{1m}) - i(v_{2n} + v_{1m})h(x)] dx \right.$$
$$\left. - \left(\frac{w_{2n}}{v_{1m}} + \frac{\mu_2}{\mu_1} \frac{w_{1m}}{v_{1m}} \right) \frac{1}{2b} \int_{-b/2}^{b/2} \exp[ix(w_{2n} - w_{1m}) - i(v_{2n} + v_{1m})h(x)]h'(x) dx \right\}, \tag{5.91}$$

where

$$f_{0n} = \frac{1}{4} \int_{-b/2}^{b/2} \left\{ \left[v_{2n} + w_{2n} h'(x) \right] U^0(x,y) - i\frac{\mu_2}{\mu_1} \left[\frac{\partial U^0}{\partial y} - \frac{\partial U^0}{\partial x} h'(x) \right] \right\} \bigg|_{y=h(x)}$$
$$\times \exp[i w_{2n} x - i v_{2n} h(x)] dx.$$

We integrate the second terms in (5.90) and (5.91) by parts and obtain

$$f_{1n} = \sum_{m=-\infty}^{\infty} f_{2m} \left[\left(\frac{v_{1n}}{v_{2m}} - \frac{\mu_1}{\mu_2} \right) + \left(\frac{w_{1n}}{v_{2m}} + \frac{\mu_1 w_{2m}}{\mu_2 v_{2m}} \right) \left(\frac{w_{1n} - w_{2m}}{v_{1n} + v_{2m}} \right) \right] h_{nm}^1, \quad (5.92)$$

$$f_{2n} = f_{0n} + \sum_{m=-\infty}^{\infty} f_{1m} \left[\left(\frac{v_{2n}}{v_{1m}} - \frac{\mu_2}{\mu_1} \right) + \left(\frac{w_{2n}}{v_{1m}} + \frac{\mu_2 w_{1m}}{\mu_1 v_{1m}} \right) \left(\frac{w_{2n} - w_{1m}}{v_{2n} + v_{1m}} \right) \right] h_{nm}^2, \quad (5.93)$$

$$f_{0n} = \frac{b}{8\pi} \left[\left(v_{2n} + \frac{\mu_2}{\mu_1} v_{10} \right) + pn \left(\frac{w_{2n} + \frac{\mu_2}{\mu_1} w_{10}}{v_{2n} - v_{10}} \right) \right] \int_{-\pi}^{\pi} \exp \left[i(v_{10} - v_{2n}) h \left(\frac{t}{p} \right) \right] e^{int} \, dt, \quad (5.94)$$

and $p = 2\pi/b$,

$$h_{nm}^1 = \frac{1}{4\pi} \int_{-\pi}^{\pi} \exp \left[i(v_{1n} + v_{2m}) h \left(\frac{t}{p} \right) \right] e^{i(n-m)t} dt, \quad (5.95)$$

$$h_{nm}^2 = \frac{1}{4\pi} \int_{-\pi}^{\pi} \exp \left[-i(v_{2n} + v_{1m}) h \left(\frac{t}{p} \right) \right] e^{i(n-m)t} dt. \quad (5.96)$$

Here it was taken into account that, according to (5.85) and (5.86),

$$w_{1n} - w_{2m} = w_{2n} - w_{1m} = p(n - m), \quad w_{2n} - k_1 \sin \theta_1 = pn,$$

and the variable of integration was changed as $px = t$. For $h(x) = \text{const}$, the second terms in the square brackets in (5.92)–(5.94) disappear.

We introduce the notation:

$$\alpha_{nm}^1 = \left[\left(\frac{v_{1n}}{v_{2m}} - \frac{\mu_1}{\mu_2} \right) + \left(\frac{w_{1n}}{v_{2m}} + \frac{\mu_1 w_{2m}}{\mu_2 v_{2m}} \right) \left(\frac{w_{1n} - w_{2m}}{v_{1n} + v_{2m}} \right) \right], \quad H_{nm}^1 = \alpha_{nm}^1 h_{nm}^1,$$

$$\alpha_{nm}^2 = \left[\left(\frac{v_{2n}}{v_{1m}} - \frac{\mu_2}{\mu_1} \right) + \left(\frac{w_{2n}}{v_{1m}} + \frac{\mu_2 w_{1m}}{\mu_1 v_{1m}} \right) \left(\frac{w_{2n} - w_{1m}}{v_{2n} + v_{1m}} \right) \right], \quad H_{nm}^2 = \alpha_{nm}^2 h_{nm}^2. \quad (5.97)$$

As a result, system (5.92), (5.93) finally becomes

$$f_{1n} = \sum_{m=-\infty}^{\infty} F_{nm}^1 f_{2m}, f_{2n} = f_{0n} + \sum_{m=-\infty}^{\infty} F_{nm}^2 f_{1m}. \quad (5.98)$$

We study the asymptotics of matrix elements and constant terms in system (5.98) for $|n|, |m| \gg 1$. As $|n| \to \infty$, $w_{jn} \sim pn$, $v_{jn} \sim -ip|n|$, and therefore, for $|n| \gg |m|$, it follows from (5.95) that

$$h_{nm}^1 \sim \frac{1}{4\pi} \int_{-\pi}^{\pi} \exp \left[-imt + iv_{2m} h \left(\frac{t}{p} \right) \right] e^{|n| \Phi_n(t)} dt, \quad (5.99)$$

where

$$\Phi_n(t) = ph\left(\frac{t}{p}\right) + is_n t, \quad s_n = \frac{n}{|n|}.$$

Now we estimate the integral (5.99) by the saddle point method; the saddle point can be obtained from the equation

$$h'\left(\frac{t_0}{p}\right) = -is_n. \tag{5.100}$$

We perform the estimate, take (5.97) into account, and obtain for $|n| \gg |m|$

$$|H_{nm}^1| \leq \frac{\text{const}}{|n|^{1/2}} \exp\left[\frac{2\sigma_1}{k} p|n|\right], \tag{5.101}$$

where

$$\sigma_1 = \frac{k}{2p} \max_{t_0, s_n} Re\Phi_n(t_0), \tag{5.102}$$

and t_0 are the roots of equation (5.100) associated with the singularities of the scattered field U^1, which are located under the curve $y = h(x)$ (see Section 1.2.2; also see [50]). Similarly, we have

$$h_{nm}^2 \sim \frac{1}{4\pi} \int_{-\pi}^{\pi} \exp\left[-imt - iv_{1m} h\left(\frac{t}{p}\right)\right] e^{-|n|\Phi_{-n}(t)} dt,$$

where

$$\Phi_{-n}(t) = ph\left(\frac{t}{p}\right) - is_n t.$$

In this case, the equation of the saddle point becomes

$$h'\left(\frac{t_0}{p}\right) = is_n. \tag{5.100a}$$

Taking (5.97) into account, we obtain

$$|H_{nm}^2| \leq \text{const}|n|^{1/2} \exp\left[-\frac{2\sigma_2}{k} p|n|\right], \tag{5.103}$$

where

$$\sigma_2 = \frac{k}{2p} \min_{t_0, s_n} Re\Phi_{-n}(t_0), \tag{5.104}$$

and t_0 are the roots of Equation (5.100) associated with the singularities of the field U_2, which are located above the curve $y = h(x)$. We note that if the function $h(t/p)$ in relations (5.102) and (5.104) has singular points, then the quantities t_0 are also understood as the coordinates of these points.

Similarly, for $|m| \gg |n|$, we obtain

$$|H_{nm}^2| \leq \frac{\text{const}}{|m|^{1/2}} \exp\left[-\frac{2\sigma_2}{k} p|m|\right]. \tag{5.105}$$

When calculating the asymptotics of the coefficients H_{nm}^1, we must use the representation

$$H_{nm}^1 = \frac{1}{4\pi} \int_{-\pi}^{\pi} \left[\left(\frac{v_{1n}}{v_{2m}} - \frac{\mu_1}{\mu_2}\right) - \left(\frac{w_{1n}}{v_{2m}} + \frac{\mu_1}{\mu_2}\frac{w_{2m}}{v_{2m}}\right) h'\left(\frac{t}{p}\right)\right] \exp\left[ih\left(\frac{t}{p}\right)(v_{1n} + v_{2m})\right] e^{i(n-m)t} dt.$$

As a result, for $|m| \gg |n|$, taking into account the fact that the integrand is zero at the saddle point, we obtain

$$|H_{nm}^1| \leq \frac{\text{const}}{|m|^{3/2}} \exp\left[\frac{2\sigma_1}{k} p|m|\right]. \tag{5.106}$$

To find the asymptotics of f_{0n} for $|n| \gg 1$, it is expedient to use the relation

$$f_{0n} = \frac{1}{4p} \int_{-\pi}^{\pi} \left[\left(v_{2n} + \frac{\mu_2}{\mu_1} v_{10}\right) + \left(w_{2n} + \frac{\mu_2}{\mu_1} w_{10}\right) h'\left(\frac{t}{p}\right)\right] \exp\left[i(v_{10} - v_{2n})h\left(\frac{t}{p}\right)\right] e^{int} dt,$$

because here the integrand is again zero at the saddle point. Estimating the integral, we obtain

$$|f_{0n}| \leq \frac{\text{const}}{|n|^{1/2}} \exp\left[-\frac{2\sigma_2}{k} p|n|\right]. \tag{5.107}$$

The quantity $2\sigma_1/k$ is equal to the maximal ordinate of the singular points of the analytic deformation of the boundary $y = h(x)$ toward the region under the boundary; similarly, $2\sigma_2/k$ is the minimal ordinate of the singular points appearing in the analytic deformation of the boundary toward the region above it.

We eliminate f_{10} and f_{20} in system (5.98) and pass to new unknown coefficients by using changes of variables, for example, of the form

$$f_{1m} = |m|^{1/2} \exp\left[\frac{2\sigma_1}{k} p|m|\right] x_{1m}, \quad f_{2m} = |m|^{3/2} \exp\left[-\frac{2\sigma_2}{k} p|m|\right] x_{2m}, \quad m \neq 0. \tag{5.108}$$

As a result, we obtain the new system

$$
x_{1n} = z_{1n}^0 + \sum_{m=-\infty}^{\infty} (1 - \delta_{m0})\left(\beta_{nm}^{11} x_{1m} + \beta_{nm}^{12} x_{2m}\right),
$$

$$
x_{2n} = z_{2n}^0 + \sum_{m=-\infty}^{\infty} (1 - \delta_{m0})\left(\beta_{nm}^{21} x_{1m} + \beta_{nm}^{22} x_{2m}\right), \quad n = \pm 1, \pm 2, \ldots
$$

(5.109)

where

$$
x_{1n}^0 = \frac{f_{00} H_{n0}^1 \exp\left[-\frac{2\sigma_1}{k} p|n|\right]}{(1 - H_{00}^1 H_{00}^2)|n|^{1/2}},
$$

$$
x_{2n}^0 = \left(f_{0n} + \frac{f_{00} H_{00}^1 H_{n0}^2}{1 - H_{00}^1 H_{00}^2}\right) \frac{\exp\left[\frac{2\sigma_2}{k} p|n|\right]}{|n|^{3/2}},
$$

$$
\beta_{nm}^{11} = \frac{H_{n0}^1 H_{0m}^2}{1 - H_{00}^1 H_{00}^2} \left|\frac{m}{n}\right|^{1/2} \exp\left[-\frac{2\sigma_1}{k} p(|n| - |m|)\right],
$$

$$
\beta_{nm}^{12} = \left(H_{nm}^1 + \frac{H_{00}^2 H_{n0}^1 H_{0m}^1}{1 - H_{00}^1 H_{00}^2}\right) \frac{|m|^{3/2}}{|n|^{1/2}} \exp\left[-\left(\frac{2\sigma_1}{k}|n| + \frac{2\sigma_2}{k}|m|\right)p\right],
$$

$$
\beta_{nm}^{21} = \left(H_{nm}^2 + \frac{H_{00}^1 H_{n0}^2 H_{0m}^2}{1 - H_{00}^1 H_{00}^2}\right) \frac{|m|^{1/2}}{|n|^{3/2}} \exp\left[\left(\frac{2\sigma_2}{k}|n| + \frac{2\sigma_1}{k}|m|\right)p\right],
$$

$$
\beta_{nm}^{22} = \frac{H_{n0}^2 H_{0m}^1}{1 - H_{00}^1 H_{00}^2} \left|\frac{m}{n}\right|^{3/2} \exp\left[\frac{2\sigma_2}{k} p(|n| - |m|)\right].
$$

(5.110)

Now it follows from estimates (5.101), (5.103), (5.105)–(5.107) and relations (5.110) that system (5.109) can be solved by the reduction method for

$$
\sigma_2 > \sigma_1.
$$

(5.111)

In the case where $\sigma_1 = -\infty$ (or $\sigma_2 = +\infty$), we must substitute any number satisfying (5.111) instead of σ_1 (respectively, σ_2) in (5.111). Condition (5.111) means that, as follows from the above, the singularities of the analytic deformation of the boundary toward the region above the surface must lie above the singularities of the analytic deformation toward the region under the surface. As noted in Ref. [51], criterion (5.111) holds for a wide class of periodic surfaces which is significantly wider than, in particular, the class of surfaces satisfying the Rayleigh hypothesis.

Now we consider several simple special cases.

1. For $h(x) = 0$, we have

$$\alpha_{nm}^1 = \left(\frac{v_{1n}}{v_{2m}} - \frac{\mu_1}{\mu_2} \right), \quad \alpha_{nm}^2 = \left(\frac{v_{2n}}{v_{1m}} - \frac{\mu_2}{\mu_1} \right), \quad h_{nm}^1 = h_{nm}^2 = \frac{1}{2} \delta_{nm}.$$

As a result, system (5.98) becomes

$$f_{1n} = \frac{1}{2} \left(\frac{v_{1n}}{v_{2n}} - \frac{\mu_1}{\mu_2} \right) f_{2n}, \quad f_{2n} = f_{0n} + \frac{1}{2} \left(\frac{v_{2n}}{v_{1n}} - \frac{\mu_2}{\mu_1} \right) f_{1n}, \tag{5.112}$$

and (see (5.94))

$$f_{0n} = \frac{b}{4} \left(v_{20} + \frac{\mu_2}{\mu_1} v_{10} \right) \delta_{n0},$$

where b is an arbitrary fixed number. Thus, from (5.98) and (5.112), we obtain

$$f_{2n} = b \frac{\mu_2 v_{10} v_{20}}{(\mu_1 v_{20} + \mu_2 v_{10})} \delta_{n0}, \quad f_{1n} = \frac{b v_{10} (\mu_2 v_{10} - \mu_1 v_{20})}{2 \ (\mu_1 v_{20} + \mu_2 v_{10})} \delta_{n0}.$$

Now we finally have for the reflection R and transmission T coefficients:

$$R \equiv \frac{2 f_{10}}{b v_{10}} = \frac{Z_2 \cos \theta_1 - Z_1 \cos \theta_2}{Z_2 \cos \theta_1 + Z_1 \cos \theta_2}, \quad Z_j = \sqrt{\mu_j / \varepsilon_j},$$

$$T \equiv \frac{2 f_{20}}{b v_{20}} = \frac{2 Z_2 \cos \theta_1}{Z_2 \cos \theta_1 + Z_1 \cos \theta_2},$$

i.e., we obtain the well-known relations.

2. Let $\varepsilon_1 = \varepsilon_2, \mu_1 = \mu_2$. Then we have

$$w_{1n} = w_{2n}, v_{1n} = v_{2n},$$

and, therefore, because $w_{1n}^2 + v_{1n}^2 = w_{2n}^2 + v_{2n}^2$ (see (5.87)), it follows from (5.97) that $\alpha_{nm}^1 = \alpha_{nm}^2 = 0$. Further, $f_{1n} = 0, f_{2n} = f_{0n}$, and $f_{0n} = \left(\frac{k_1 b}{2} \cos \theta_1 \right) \delta_{n0}$, i.e.,

$$U^1 = 0, \quad U_2 = \frac{2}{b} \left(\frac{k_1 b}{2} \cos \theta_1 \right) \frac{e^{-ik_1 x \sin \theta_1 + ik_1 y \cos \theta_1}}{k_1 \cos \theta_1} = U^0.$$

3. Now let $h(x) = a \cos px$, $a > 0$. For such a geometry, from (5.94) and (5.97) we obtain

$$h_{nm}^1 = \frac{i^{n-m}}{2} J_{n-m}(a(v_{1n} + v_{2m})), \quad h_{nm}^2 = \frac{i^{n-m}}{2} J_{n-m}(a(v_{2n} + v_{1m})),$$

$$f_{0n} = \alpha_{0n} \frac{i^n}{2} J_n(a(v_{10} - v_{2n})),$$

where

$$\alpha_{0n} = \frac{b}{2}\left[\left(v_{2n} + \frac{\mu_2}{\mu_1}v_{10}\right) + pn\left(\frac{w_{2n} + \frac{\mu_2}{\mu_1}w_{10}}{v_{2n} - v_{10}}\right)\right]$$

In these examples, the calculations of the matrix elements and constant terms in system (5.98) are comparatively simple. In the general case, it is necessary to use relations (5.94)–(5.97).

The above approach can obviously be generalized to problems with two boundaries, i.e., to problems of a plane wave scattering at a layer of finite thickness such that one of its boundaries is a periodic surface and the other can be either periodic or plane. The method can also be used to determine the natural waves in periodic structures, in particular, in the so-called open periodic waveguides.

5.6 CALCULATION OF THE REFLECTION AND TRANSMISSION COEFFICIENTS IN A PLANE DIELECTRIC WAVEGUIDE WITH FOREIGN OBJECTS NEAR IT

In this section, we use the pattern equation method to solve the problem of wave propagation in a dielectric waveguide [107]. We introduce Cartesian coordinates as shown in Fig. 5.17. For definiteness, we consider the case of diffraction of a TE-wave propagating in a dielectric layer. Thus, in the considered geometry, the vector $\vec{E}$ is directed along the generating element of the cylindrical body, i.e., has a unique component $U = E_z(x, y)$. The case of a different polarization can be considered similarly.

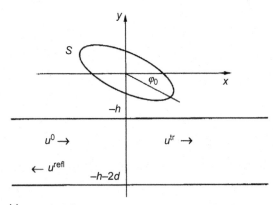

Figure 5.17 Problem geometry.

We also assume that the Dirichlet condition is satisfied on the body surface and the usual conjugation conditions are satisfied on the interface between the media ($y = -h$ and $y = -h - 2d$ in Fig. 5.17) for the tangent components of the magnetic and electric fields.

To solve this problem, we represent the diffraction field as an integral of plane waves [102,109]. This representation can be obtained by using the Green function, which takes account of the shape of the plane layered media contained in the scatterer. For the model under study, the Green function has the form

$$G(\vec{r}, \vec{r}') = -\frac{i}{4}H_0^{(2)}\left(k|\vec{r} - \vec{r}'|\right) + G_{\text{reg}}(\vec{r}, \vec{r}').\tag{5.113}$$

where

$$G_{\text{reg}}(\vec{r}, \vec{r}') = -\frac{i}{4\pi}\int_{-\infty}^{\infty} R(\kappa)e^{-i2\gamma kh - i\gamma k(y+y')}e^{-i\kappa k(x-x')}\frac{d\kappa}{\gamma}.\tag{5.114}$$

In formula (5.114), $\gamma = \sqrt{1 - \kappa^2}$ $(\text{Im}\gamma \leq 0)$ and k is the wave number of the dielectric plane environment. The function $R(\kappa)$ has the form

$$R(\kappa) = \frac{R_0(\kappa)[\exp(-i4\gamma kd) - 1]}{1 - R_0^2(\kappa)\exp(-i4\gamma kd)},\tag{5.115}$$

where

$$R_0(\kappa) = \frac{\sqrt{n^2 - \kappa^2} - \gamma}{\sqrt{n^2 - \kappa^2} + \gamma},\tag{5.116}$$

and n is the relative refraction index of the layer medium and the environment. We note that the function $R(\kappa)$ is the coefficient of the plane wave reflection from the dielectric layer. According to formula (1.9), the diffraction field everywhere outside the domain of the cylindrical obstacle can be expressed as the integral

$$\int_S \left\{ U(\vec{r}')\frac{\partial G_0(\vec{r}; \vec{r}')}{\partial n'} - \frac{\partial U(\vec{r}')}{\partial n'}G_0(\vec{r}; \vec{r}')\right\}ds' = U(\vec{r}) - U^0(\vec{r}).$$

$$\tag{5.117}$$

In formula (5.117), the differentiation is performed along the outer normal to the body boundary. The function U^0 is the field of the principal mode of the dielectric layer, which has the form

$$U^0(\vec{r}) = \cos(\gamma_0 d)\exp[-\alpha_0 k(y+h) - i\kappa_0 kx], \quad y > -h. \tag{5.118}$$

In formula (5.118), we use the notation

$$\gamma_0 = \sqrt{n^2 - \cos^2\psi_0}, \quad \alpha_0 = \sqrt{(n^2 - 1) - \gamma_0^2}, \tag{5.119}$$

where ψ_0 is the first root of the transcendental equation $tg\gamma kd = \alpha/\gamma$ [31].

Substituting the expression for the Green function in the form of relations (5.113)–(5.116) into formula (5.117), we can write the diffraction field in the half-space $y > -h$ as follows (after the change of variables $\kappa = \cos\psi, \gamma = \sin\psi$):

$$U^1 = U_c^1 + U_{\text{reg}}^1 \tag{5.120}$$

$$U_c^1(r, \varphi) = \frac{1}{\pi} \int_{-(\pi/2)-i\infty}^{\pi/2+i\infty} g(\psi + \varphi)e^{-ikr\cos\psi}\,d\psi, \tag{5.121}$$

$$U_{\text{reg}}^1(r, \varphi) = \frac{1}{\pi} \int_{-i\infty}^{\pi+i\infty} g(-\psi)R(-\psi)e^{-i2kh\sin\psi} \cdot e^{-ikr\cos(\psi-\varphi)}\,d\psi. \tag{5.122}$$

In the last formula (5.122), we let $R(-\psi)$ denote the function $R(\kappa)$ after this change. In formulas (5.121) and (5.122), $g(\alpha)$ denotes the standard spectral function of the wave field, which has the form

$$g(\alpha) = \int_0^{2\pi} \left(\hat{D}U\right) e^{ik\rho(\varphi)\cos(\alpha-\varphi)}\,d\varphi, \tag{5.123}$$

where the operator

$$\hat{D}U = \frac{i}{4}\left[\rho(\varphi)\frac{\partial U}{\partial r} - \frac{\rho'(\varphi)}{\rho(\varphi)}\frac{\partial U}{\partial r}\right]_{r=\rho(\varphi)}. \tag{5.124}$$

Here $r = \rho(\varphi)$ is the equation of the contour of the body cross-section in cylindrical coordinates. We note that relation (5.121) is the part of the scattered field generated by the "currents" distributed over the body surface, and the value of $U_{pe\partial}^1$ is determined by the field of plane waves reflected from the dielectric layer boundary. Further, we derive the integral equation for the spectral function. For this, we substitute the sum of the initial and diffraction fields into formulas (5.123) and (5.124). Then, after simple transformations with formulas (5.120)–(5.122) taken into account, we obtain [102,103]

$$g(\alpha) = g^0(\alpha) + \int_\Gamma K_0(\alpha, \psi)g(\psi)\,d\psi + \int_{\Gamma'} K_1(\alpha, \psi)g(-\psi)\,d\psi, \tag{5.125}$$

where the contours Γ and Γ' coincide with the contours in formulas (5.121) and (5.122), respectively. The kernels $K_0(\alpha, \psi)$ and $K_1(\alpha, \psi)$ have the form

$$K_0(\alpha, \psi) = \frac{k}{4\pi} \int_0^{2\pi} [\rho(\varphi)\cos\psi - \rho'(\varphi)\sin\psi]$$

$$\times \exp\left\{-ik\rho(\varphi)(\cos\psi - \cos(\alpha - \varphi)) + \varphi\frac{d}{d\psi}\right\} d\varphi, \tag{5.126}$$

$$K_1(\alpha, \psi) = \frac{k}{4\pi} \int_0^{2\pi} R(\psi) e^{-i2kh\sin\psi}$$

$$\times \hat{D}\{\exp[-ikr\cos(\psi - \varphi)]\} e^{ik\rho(\varphi)\cos(\alpha-\varphi)} d\varphi. \tag{5.127}$$

The function $g^0(\alpha)$ in Equation (5.125) is the "pattern" of the initial field, which can be determined by the formula

$$g^0(\alpha) = \int_0^{2\pi} \left(\hat{D} U^0\right) e^{ik\rho(\varphi)\cos(\alpha-\varphi)} d\varphi. \tag{5.128}$$

Now let us algebraize Equation (5.125). We represent the functions $g^0(\alpha)$ and $g(\alpha)$ as the Fourier series

$$g(\alpha) = \sum_{m=-\infty}^{\infty} a_m e^{im\alpha}, \quad g^0(\alpha) = \sum_{m=-\infty}^{\infty} a_m^0 e^{im\alpha}. \tag{5.129}$$

Then, substituting the series (5.129) in Equation (5.125), we obtain the algebraic system for the unknown coefficients a_m,

$$a_m = a_m^0 + \sum_{n=-\infty}^{\infty} G_{mn} a_n. \tag{5.130}$$

In this system,

$$G_{mn} = G_{mn}^\infty + \sum_{p=-\infty}^{\infty} G_{mp}^0 S_{pn}, \tag{5.131}$$

where

$$G_{mn}^\infty = \frac{1}{4} \int_0^{2\pi} J_m(k\rho) \left[ik\rho(\varphi) H_n^{(2)\prime}(k\rho) + n\frac{\rho'(\varphi)}{\rho(\varphi)} H_n^{(2)}(k\rho) \right]$$

$$\times \exp\left[i(n-m)\left(\varphi - \frac{\pi}{2}\right) \right] d\varphi \tag{5.132}$$

are elements of the matrix corresponding to the problem of wave scattering at the considered body in a homogeneous medium and

$$G^0_{mn} = \frac{1}{4} \int_0^{2\pi} J_m(k\rho) \left[ik\rho(\varphi) J_n^{(2)\prime}(k\rho) + n \frac{\rho'(\varphi)}{\rho(\varphi)} J_n^{(2)}(k\rho) \right]$$
$$\times \exp\left[i(n-m)\left(\varphi - \frac{\pi}{2}\right) \right] d\varphi \tag{5.133}$$

are elements of the matrix, which is determined only by the scatterer geometry and has the same form in different problems [103]. Finally,

$$S_{mn} = \frac{1}{\pi} \int_\Gamma R(-\psi) e^{-i2kh\sin\psi} e^{-i(m+n)\psi} d\psi \equiv S_{m+n}. \tag{5.134}$$

To calculate S_{m+n} efficiently, it is necessary to transform expression (5.134) (see [107]).

After simple transformations, we obtain the expression

$$d^0_m = \frac{i}{4} A_0 e^{-im\varphi_0} \int_0^{2\pi} \left[k\rho'(\varphi) \left(\cos(\varphi + \varphi_0) \sqrt{\kappa_0^2 - 1} - i\kappa_0 \sin(\varphi + \varphi_0) \right) \right.$$
$$\left. - k_1\rho(\varphi) \left(\sin(\varphi + \varphi_0) \sqrt{\kappa_0^2 - 1} + i\kappa_0 \cos(\varphi + \varphi_0) \right) \right] \cdot e^{-k_1\rho(\varphi)\sin(\varphi + \varphi_0)\sqrt{\kappa_0^2 - 1}}$$
$$\times e^{-i\kappa_0 k\rho(\varphi)\cos(\varphi + \varphi_0)} J_m(k\rho(\varphi)) e^{-im(\varphi - \pi/2)} d\varphi, \tag{5.135}$$

for the right-hand side of SLAE (5.130).

Here

$$A_0 = \cos\left(kd\sqrt{n^2 - \kappa_0^2} \cdot \exp(-\alpha_0 kh) \right), \tag{5.136}$$

and φ_0 is the angle of the body rotation with respect to the horizontal.

5.6.1 Asymptotic Solution

We introduce the notation

$$L[g(\alpha)] \equiv g(\alpha) - \int_\Gamma K_0(\alpha, \psi) g(\psi) d\psi$$

and rewrite Equation (5.125) as

$$L[g(\alpha)] = A_0 g^0_{un}(\alpha; \psi_0) + \int_{\Gamma'} K_1(\alpha, \psi) g(-\psi) d\psi, \tag{5.137}$$

where g^0_{un} is the function g^0 (see (5.128)) after division by the multiplier A_0. If we let $g^\infty_{un}(\alpha; \psi_0)$ denote the scattering pattern of the body in the homogeneous medium after the incidence of the unit plane wave propagating at an angle ψ_0, then it follows from (5.137) that this formula can be obtained from the equation

$$L[g^\infty(\alpha; \psi_0)] = g_{un}^0(\alpha; \psi_0).$$

Now Equation (5.137) becomes (see (5.127))

$$L[g(\alpha)] = A_0 L[g^\infty(\alpha; \psi_0)] + \frac{1}{\pi} \int_{\Gamma'} L[g^\infty(\alpha; \psi)] R(-\psi) g(-\psi) e^{-i2kh\sin\psi} d\psi,$$

whence, by the linearity of the operator L, we obtain the Tversky-type Fredholm integral equation of the second type (e.g., see [139]) for the function $g(\alpha)$:

$$g(\alpha) = A_0 g^\infty(\alpha; \psi_0) + \frac{1}{\pi} \int_{-i\infty}^{\pi+i\infty} g^\infty(\alpha; \psi) R(-\psi) g(-\psi) e^{-i2kh\sin\psi} d\psi.$$

$$(5.138)$$

Equation (5.138) can be solved analytically for $2kh \gg 1$. More precisely, this condition can be written as $h > kD\,(D/2\pi)$, where D is the maximal cross-section of the object. This solution can be obtained by estimating the integral in (5.138) by the saddle point method. Figure 5.18 shows the

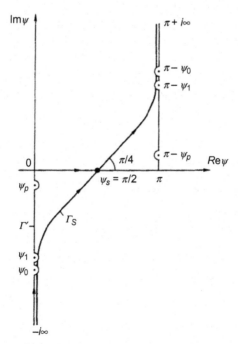

Figure 5.18 Contours of integration.

initial contour of integration Γ' and the saddle point contour Γ_S and illustrates the location of the integrand poles.

Estimating the integral in (5.138) by the saddle point method technique, we obtain

$$g(\alpha) \approx A(\psi_0)g^\infty(\alpha; \psi_0) + \frac{\exp(-i2kh + i\pi/4)}{\sqrt{\pi kh}}g^\infty\left(\alpha; \frac{\pi}{2}\right)g\left(-\frac{\pi}{2}\right)R\left(-\frac{\pi}{2}\right).$$

$$(5.139)$$

It follows from (5.139) that

$$g\left(-\frac{\pi}{2}\right) = \frac{A(\psi_0)g^\infty\left(-\frac{\pi}{2};\psi_0\right)}{1 - \dfrac{\exp(-i2kh + i\pi/4)}{\sqrt{\pi kh}}g^\infty\left(-\frac{\pi}{2};\frac{\pi}{2}\right)R\left(-\frac{\pi}{2}\right)}, \qquad (5.140)$$

and it follows from (5.115) and (5.116) that

$$R\left(-\frac{\pi}{2}\right) = \frac{i(1 - n^2)}{2nctg(2nkd) + i(1 + n^2)}. \qquad (5.141)$$

Relations (5.139)–(5.141) give the complete solution of the problem, i.e., all parameters of the problem can be calculated if the scattering pattern $g(\alpha)$ of the object is known.

The quantity $g^\infty(\alpha; \beta)$ required for all calculations can be determined by solving the corresponding boundary-value problem for the scatterer in the homogeneous medium, for example, by the method of auxiliary currents (see Chapter 2). In the case where the scatterer is, for example, a circular cylinder, relations (5.139)–(5.141) permit writing the solution in explicit analytic form.

The field inside the waveguide can be represented as

$$U = U^0 + U^S,$$

where

$$U^0 = \cos\left[k(y + h + d)\sqrt{n^2 - \cos^2\psi_0}\right]\exp(-ikx\cos\psi_0). \qquad (5.142)$$

For U^S, using the representation for the Green function $G(\vec{r}; \vec{r}')$ inside the layer (e.g., see [103]), we obtain

$$U^S(x, y) = \frac{1}{\pi}\int_{-i\infty}^{\pi + i\infty} \left\{\left[T(-\psi)\exp\left(ik(h + y)\sqrt{n^2 - \cos^2\psi}\right)\right.\right.$$
$$+ \tilde{R}(-\psi)\exp\left(-ik(h + 4d + y)\sqrt{n^2 - \cos^2\psi}\right)\Big]g(-\psi)\exp(-ikh\sin\psi)$$
$$\times \frac{\exp(-ikx\cos\psi)}{\sqrt{n^2 - \cos^2\psi}}\sin\psi\right\}d\psi,$$

$$(5.143)$$

where

$$T(-\psi) = \frac{T_{12}}{1 - R_{12}^2 \exp\left(-i4kd\sqrt{n^2 - \cos^2\psi}\right)};$$

$$\widetilde{R}(-\psi) = \frac{-T_{12}R_{12}}{1 - R_{12}^2 \exp\left(-i4kd\sqrt{n^2 - \cos^2\psi}\right)};$$

and in this case

$$T_{12} = \frac{2\sqrt{n^2 - \cos^2\psi}}{\sin\psi + \sqrt{n^2 - \cos^2\psi}}; \quad R_{12} = \frac{\sin\psi - \sqrt{n^2 - \cos^2\psi}}{\sin\psi + \sqrt{n^2 - \cos^2\psi}}.$$

To transform the integral in (5.143) for $x > 0$, we can use the relation

$$\frac{1}{\pi}\int_{-i\infty}^{\pi+i\infty} - \frac{1}{\pi}\int_{-(\pi/2)-i\infty}^{\pi/2+i\infty} + \frac{1}{\pi}\int_{C_\infty} \{\ldots\} d\psi = 2i\sum_m \operatorname*{res}_{\psi_m}\{\ldots\},$$

where the symbol $\{\ldots\}$ denotes the integrand in (5.143) and the symbol C_∞ denotes the integrals over the horizontal intervals connecting the point $((-\pi/2) - i\infty)$ with $(0 - i\infty)$ and the point $(\pi + i\infty)$ with $(\pi/2 + i\infty)$.

When $x > 0$, the integrals over C_∞ are equal to zero. Therefore, expression (5.143) can be rewritten for $x > 0$ as

$$U^S(x, y) = \frac{1}{\pi}\int_{-(\pi/2)-i\infty}^{\pi/2+i\infty} \{\ldots\} d\psi + 2i\sum_m \operatorname*{res}_{\psi_m}\{\ldots\}, \tag{5.144}$$

and ψ_m are the poles of the function $R(-\psi)$ corresponding to the eigenwaves of the plane dielectric waveguide, which can be determined from the transcendental equations [31]:

$$\begin{cases} tg\gamma d = \dfrac{\alpha}{\gamma}, & m = 0, 2, 4, \ldots, \\[2mm] tg\gamma d = -\dfrac{\gamma}{\alpha}, & m = 1, 3, 5, \ldots \end{cases} \tag{5.145}$$

For $kx \gg 1$, the integral in (5.144) can again be estimated by the saddle point method if we take into account the fact that the integrand is zero at the saddle point ψ_s (in this case, $\psi_s = 0$). As the result of estimation, we obtain

$$u^S(x, y) \approx \frac{F''(0)}{\sqrt{2\pi}} \frac{\exp(-ikx + i3\pi/4)}{(kx)^{3/2}} + 2i\sum_m \operatorname*{res}_{\psi_m}\{F(\psi)\exp(-ikx\cos\psi)\}, \tag{5.146}$$

where $F(\psi)$ denotes the entire integrand in (5.143) except for the multiplier $\exp(-ikx\cos\psi)$. The sum in (5.144) is the set of eigen-waves in the waveguide. Because $k\cos\psi_m = \beta_m$ is the constant of propagation of the mth eigen-wave, we set $y = -h - d$ in this sum and obtain the values of the transmission coefficients as the multipliers of $\exp(-i\beta_m x)$.

Similarly, for $x < 0$, the initial contour of integration in (5.143) can be moved to the right by $\pi/2$, which results in the appearance of residuals at the points $\pi - \psi_m$. Proceeding as above, we determine the reflection coefficients as the corresponding multipliers of $\exp(+i\beta_m x)$.

5.6.2 Results of Calculations

Figure 5.19 illustrates the dependence of the reflection R and transmission T coefficients on the parameter $F = 2kd\sqrt{n^2 - 1}$ for different dimensions of the rectangular scatterer of width $2b$ and height $2a$. The following notation is used here: curve 1 corresponds to $h/2d = 2.01$, $a/2d = 2$, $b/2d = 1$; curve 2 corresponds to $h/2d = 1.01$, $a/2d = 1$, $b/2d = 1$; and curve 3 corresponds to $h/2d = 2.01$, $a/2d = 1$, $b/2d = 1$.

Figure 5.20 shows the dependence of the reflection and transmission coefficients of the principal wave mode of a plane dielectric waveguide on the parameter F calculated by rigorous and asymptotic relations in the case where the scatterer is an ideally conducting circular cylinder of radius $a/2d = 1$ (curves 1 and 2) and $a/2d = 0.5$ (curves 3). The solid curves present the exact solution, and the dash-dotted curves correspond to the asymptotic solution. Here curve 1 corresponds to $h/2d = 1.1$, $a/2d = 1$; curve 2 corresponds to $h/2d = 1.01$, $a/2d = 1$; and curve 3 corresponds to $h/2d = 0.51$,

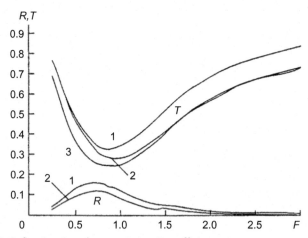

Figure 5.19 Reflection R and transmission T coefficients.

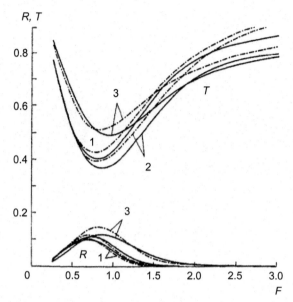

Figure 5.20 Reflection R and transmission T coefficients calculated by rigorous and asymptotic relations.

$a/2d = 0.5$. One can see that the results of calculations by rigorous and asymptotic relations are sufficiently close to each other.

In conclusion, we note that *a priori* information about the singularities of the analytic continuation of wave fields is also extremely important when the inverse problems of the theory of diffraction and antennas are solved (e.g., see [9,10,144–147]), but this field of study is beyond the scope of this book.

REFERENCES

[1] Hönl H, Maue A, Westpfal K. Theorie der Beugung. Berlin/Heidelberg: Springer-Verlag; 1961.

[2] Kyurkchan AG. Influence of the edge effects of the radiation pattern of antenna array. Radio Eng Electron Phys 1980;25(4):22–6.

[3] Medgyesi-Mitschang LN, Wang D-S. Hybrid solutions for large-impedance coated bodies of revolution. IEEE Trans Ant Prop 1986;AP-34(11):1319–29.

[4] Tikhonov AN, Samarskii AA. Equations of mathematical physics. Moscow: Nauka; 1972 [in Russian].

[5] Colton D, Kress R. Integral equation methods in scattering theory. New York: John Wiley; 1984.

[6] Courant R. Partial differential equations. New York: Interscience; 1965.

[7] Papas CH. Theory of electromagnetic wave propagation. New York/London/Sydney: McGraw-Hill Book Comp.; 1965.

[8] Shenderov EL. Radiation and scattering of sound. Leningrad: Sudostroenie; 1989 [in Russian].

[9] Apel'tsin VF, Kyurkchan AG. Analytic properties of wave fields. Moscow: Moscow State University; 1990 [in Russian].

[10] Kyurkchan AG, Sternin BYu, Shatalov VYe. The singularities of the continuation of wave fields. Phys Usp 1996;39:1221–42.

[11] Zakharov EV, Pimenov YuV. Numerical analysis of diffraction of radio waves. Moscow: Radio and Communication; 1982 [in Russian].

[12] Vasiliev EN. Excitation of bodies of revolution. Moscow: Radio and Communication; 1987 [in Russian].

[13] Kyurkchan AG, Anyutin AP. The method of extended boundary conditions and wavelets. Dokl Math 2002;66(1):132–5.

[14] Kyurkchan AG, Smirnova NI. Solution of diffraction problems using the methods of extended boundary conditions and discrete sources. J Commun Technol Electron 2005;50(10):1139–46.

[15] Kravtsov VV. The approximation of the electromagnetic field of the antenna potential. Dokl Akad Nauk SSSR 1979;248(2):328–31 [in Russian].

[16] Kyurkchan AG. On the method of auxiliary currents and sources in problems of wave diffraction. Radio Eng Electron Phys 1984;29(10–11):2129–39.

[17] Kupradze VD. About approximate solution of mathematical physics problems. Usp Math Nauk 1967;22(2):58–109 [in Russian].

[18] Aleksidze MA. Fundamental functions in approximate solutions of the boundary problems. Moscow: Nauka; 1991 [in Russian].

[19] Waterman PC. Matrix formulation of electromagnetic scattering. Proc IEEE 1965;53:805–12.

[20] Mishchenko MI, Videen G, Babenko VA, Khlebtsov NG, Wriedt T. T-matrix theory of electromagnetic scattering by particles and its applications: a comprehensive reference database. J Quant Spectrosc Radiat Transf 2004;88:357–72.

[21] Kyurkchan AG, Smirnova NI. Solution of wave diffraction problems on impedance scatterers by the method of continued boundary conditions. J Quant Spectrosc Radiat Transf 2007;106(1–3):203–11.

[22] Kyurkchan AG, Smirnova NI. Generalization of the method of extended boundary conditions. J Commun Technol Electron 2008;53(7):767–74.

[23] Kyurkchan AG, Smirnova NI. Solution of diffraction problems by null field and T-matrix methods with accounting for wave field analytical continuation singularities, In: Proceedings of the International Conference "Days on Diffraction 2009". May 26–29; 2009. p. 133–9.

[24] Eremin YuA. Representation of fields in terms of sources in complex plane in the method of non-orthogonal series. Sov Phys Dokl 1983;28(6):451–2.

[25] Wriedt T, editor. Generalized multipole techniques for electromagnetic and light scattering. Amsterdam: Elsevier; 1999.

[26] Kyurkchan AG, Minaev SA, Soloveichik AL. A modification of the method of discrete sources based on a priori information about the singularities of the diffracted field. J Commun Technol Electron 2001;46(6):615–21.

[27] Doicu A, Eremin Y, Wriedt T. Acoustic and electromagnetic scattering analysis using discrete sources. London: Academic Press; 2000.

[28] Doicu A, Wriedt T, Eremin Y. Light scattering by systems of particles. Null-field method with discrete sources—theory and programs. Berlin/Heidelberg/New York: Springer; 2006.

[29] Wriedt T. Review of the null-field method with discrete sources. J Quant Spectrosc Radiat Transf 2007;106:535–45.

[30] Kyurkchan AG. Representation of diffraction fields by wave potentials and the method of auxiliary currents in problems of diffraction of electromagnetic waves. Sov J Commun Technol Electron 1986;31(5):20–7.

[31] Fel'd YaN, Benenson LS. Fundamentals of antenna theory. Moscow: Drofa; 2007 [in Russian].

[32] Pimenov YuV, Vol'man VI, Muravtsov AD. Technical electrodynamics. Moscow: Radio and Communication; 1982 [in Russian].

[33] Zommerfeld A. Partielle differentialgleichungen der physik. Leipzig: Akademische Verlagsgesellschaft Geest and Portig; 1948.

[34] Kyurkchan AG. Solving equations of mathematical physics by reducing them to ordinary differential equations with operator coefficients. Sov Phys Dokl 1992;37:70–5.

[35] Olver FWJ. Asymptotics and special functions. New York/London: Academic Press; 1974.

[36] Wilcox CH. An expansion theorem for electromagnetic fields. Commun Pure Appl Math 1956;9:115–34.

[37] Kyurkchan AG. On the exact value of the radius of convergence of the Wilcox series. Sov Phys Dokl 1991;36(8):567–9.

[38] Borovikov VA, Kinber BE. Geometric theory of diffraction. Moscow: Communication; 1978 [in Russian].

[39] Vilenkin NYa. Special functions and the theory of group representations. Moscow: Nauka; 1965 [in Russian].

[40] Ivanov EA. Diffraction of electromagnetic waves on two bodies. Minsk: Nauka i Tehnika; 1968 [in Russian].

[41] Kyurkchan AG. Limits of applicability of the Rayleigh and Sommerfeld representations in three-dimensional wave diffraction problems. Radio Eng Electron Phys 1983;28(7):33–41.

[42] Do Dyk Tkhang, Kyurkchan AG. Efficient method for solving the problems of wave diffraction by scatterers with broken boundaries. Acoust Phys 2003;49(1):43–50.

[43] Mittra R. Computer techniques for electromagnetics. Oxford, New York: Pergamon Press; 1973.

[44] Kyurkchan AG. On solving the problem of plane wave diffraction by gratings. Phys Dokl 1996;41(9):618–23.

[45] Kyurkchan AG. The radiation pattern of an element in diffraction grating theory and Rayleigh hypothesis. Radio Eng Electron Phys 1983;28(8):53–7.

[46] Kyurkchan AG. Methods for solving of electrodynamics boundary problems. In: Fel'd Ya N, Zelkin EG, editors. Antenna handbook. Moscow: Radiotekhnika; 1997, in Russian.

[47] Khurgin YaI, Yakovlev VP. Compactly supported functions in physics and engineering. Moscow: Nauka; 1971 [In Russian].

[48] Kyurkchan AG. Analytical continuation of wave fields. Sov J Commun Technol Electron 1986;31(11):59–69.

[49] Sternin B, Shatalov V. Differential equations on complex manifolds. Dodrecht/Boston: Academic Publishers; 1994.

[50] Kyurkchan AG. Analytic continuation of wave fields in the problem of plane wave scattering by a periodic surface. Dokl Akad Nauk SSSR 1987;292(6):1350–5 [in Russian].

[51] Kyurkchan AG. On a new class of equations in diffraction theory. J Commun Technol Electron 1993;38(6):87–98.

[52] Kyurkchan AG, Manenkov SA, Negorozhina ES. Simulation of wave scattering by a group of closely spaced bodies. J Commun Technol Electron 2008;53 (3):256–65.

[53] Kyurkchan AG. The Rayleigh and Sommerfeld diffracted field representations and the domains of their convergence. Radio Eng Electron Phys 1982;27(2):35–43.

[54] Kyurkchan AG. Analytical properties of wave fields and patterns. Moscow: IRE AN SSSR; 1984 [Preprint No. 12(384); in Russian].

[55] Wiener N. The Fourier integral and certain of its applications. Cambridge: Cambridge University press; 1933.

[56] Kyurkchan AG, Sukov AI. A method of auxiliary spline-currents for wave diffraction problems. Dokl Phys 2000;45(6):269–73.

[57] Kyurkchan AG, Minaev SA. Solution of wave diffraction problems using the wavelet technique. J Commun Technol Electron 2003;48(5):500–5.

[58] Kyurkchan AG. Appendix B. In: Blatter K, editor. Wavelet analysis. Fundamentals of the theory. Moscow: Technosphera; 2004 [in Russian].

[59] Anyutin AP, Kyurkchan AG, Minaev SA. A modified method of discrete sources. J Commun Technol Electron 2002;47(8):864–9.

[60] Anioutine AP, Kyurkchan AG, Minaev SA. About a universal modification of the method of discrete sources and its application. J Quant Spectrosc Radiat Transf 2003;79–80:509–20.

[61] Kyurkchan AG, Sukov AI. The singularities of scattered field and their role in the method of auxiliary sources. Dokl Akad Nauk SSSR 1988;303(6):1347–9 [in Russian].

[62] Apel'tsin VF, Kyurkchan AG, Sukov AI. On the role of singularities of the analytic continuation of the scattered field in the method of secondary sources. Izvestiya VUZov Radiophysics 1989;32(8):1042–6 [in Russian].

[63] Kyurkchan AG, Sukov AI, Kleev AI. The methods for solving the problems of diffraction of electromagnetic and acoustic waves using the information on analytic properties of the scattered field. ACES Journal 1994;9(3):101–12.

[64] Kyurkchan AG, Kleev AI. The use of a priori information on analytic properties of a solution in electromagnetic and acoustic problems. J Commun Technol Electron 1996;41(2):145–53.

[65] Kyurkchan AG, Kleev AI, Sukov AI. Singularities of wave fields and numerical methods of solving the boundary-value problems for Helmholtz equation. In: Wriedt T, editor. Generalized multipole techniques for electromagnetic and light scattering. Amsterdam: Elsevier; 1999.

[66] Kyurkchan AG, Kleev AI, Sukov AI. Singularities of wave fields and numerical methods for solving boundary value problems for the Helmholtz equation. Foreign electronics 2000;2:14–33.

[67] Kyurkchan AG, Anyutin AP. Wavelets and the method of auxiliary currents. Dokl Math 2002;65(2):296–300.

[68] Anyutin AP, Kyurkchan AG. Solution of diffraction problems using the auxiliary current method and wavelet bases. J Commun Technol Electron 2005;47(6):615–20.

[69] Kyurkchan AG, Manenkov SA. Analysis of electromagnetic wave scattering from an irregularity in a chiral half-space based on the modified method of discrete sources. J Commun Technol Electron 2003;48(8):821–30.

[70] Anioutine AP, Kyurkchan AG. Application of wavelets technique to the integral equations of the method of auxiliary currents. J Quant Spectrosc Radiat Transf 2003;79–80:495–508.

[71] Kyurkchan AG, Manenkov SA. The application of the modified discrete sources method to the problem of wave diffraction on a body in chiral half-space. J Quant Spectrosc Radiat Transf 2004;89:201–18.

[72] Kyurkchan AG, Minaev SA. Using of the wavelet technique for the solution of the wave diffraction problems. J Quant Spectrosc Radiat Transf 2004;89:219–36.

[73] Kyurkchan AG, Anyutin AP. The well-posedness of the formulation of diffraction problems reduced to Fredholm integral equations of the first kind with a smooth kernel. J Commun Technol Electron 2006;51(1):48–51.

[74] Anioutine AP, Kyurkchan AG, Manenkov SA, Minaev SA. About 3D solution of diffraction problems by MDS. J Quant Spectrosc Radiat Transf 2006;100:26–41.

[75] Kyurkchan AG, Manenkov SA, Negorozhina ES. Analysis of electromagnetic field diffraction by bodies of revolution by using the modified method of discrete sources. J Commun Technol Electron 2006;51(11):1209–17.

[76] Kyurkchan AG, Manenkov SA. The application of modified method of discrete sources for solving the problem of wave scattering by group of bodies. J Quant Spectrosc Radiat Transf 2008;109:1430–9.

[77] Vzyatyshev VF, Katsenelenbaum BZ, Kyurkchan AG, Manenkov SA. Diffraction by a compact obstacle near an open end of a planar dielectric waveguide. J Commun Technol Electron 2011;56(5):565–71.

[78] Kyurkchan AG, Smirnova NI. Accounting of the wave field analytic continuation singularities by using the null-field and T-matrix methods. Electromagn Waves Electron Syst 2008;13(8):78–86 [in Russian].

[79] Kyurkchan AG, Smirnova NI. Solution of wave diffraction problems by the null-field method. Acoust Phys 2009;55(6):691–7.

[80] Kyurkchan AG, Smirnova NI. Methods of auxiliary currents and the null-field. Electromagn Waves Electron Syst 2009;14(8):4–12 [in Russian].

[81] Kyurkchan AG, Manenkov SA, Smirnov VI. Solution of the problem of diffraction at a grating of bodies of revolution by the modified null field method. J Commun Technol Electron 2010;55(8):841–54.

[82] Kyurkchan AG, Manenkov SA. Analysis of diffraction of a plane wave on a grating consisting of impedance bodies of revolution. J Quant Spectrosc Radiat Transf 2011;112:1343–52.

[83] Demin DB, Kyurkchan AG, Smirnova NI. Averaging over the angles of radiation in the two-dimensional scalar problem of diffraction. T-Comm Telecommun Transport 2012;6(11):15–21 [in Russian].

[84] Anyutin AP, Kyurkchan AG. Solution of diffraction and antenna problems using the method of extended boundary conditions and wavelet technique. J Commun Technol Electron 2004;49(1):11–9.

[85] Kyurkchan AG, Manenkov SA. Electromagnetic wave diffraction at a large asperity on an impedance plane. J Commun Technol Electron 2004;49(12):1319–26.

[86] Kyurkchan AG, Smirnova NI. Solution of the electromagnetic waves diffraction problems by the method of continued boundary conditions. Antennas 2006;115(12):3–7 [in Russian].

[87] Kyurkchan AG, Smirnova NI. Solution of wave diffraction problems by the method of continued boundary conditions. Acoust Phys 2007;53(4):426–35.

[88] Anyutin AP, Kyurkchan AG, Smirnova NI. Modeling the characteristics of mirror antennas on the basis of continued boundary conditions. Electromagn Waves Electron Syst 2007;12(8):63–70 [in Russian].

[89] Tikhonov AN, Arsenin VYa. Methods for solving Ill-posed problems. Moscow: Nauka; 1986 [in Russian].

[90] Skobelev SP. Application of extended boundary conditions and the Haar wavelets in the analysis of wave scattering by thin screens. J Commun Technol Electron 2006;51 (7):748–58.

[91] Bohren KF, Huffman DR. Absorption and scattering of light by small particles. New York: John Wiley & Sons; 1983.

[92] Ufimtsev PYa. The theory of the edge diffraction waves in electrodynamics. Moscow: Binom; 2007 [in Russian].

[93] Hansen RC, editor. Microwave scanning antennas. New York/London: Academic Press; 1964.

[94] Kyurkchan AG. A new integral equation in the diffraction theory. Sov Phys Dokl 1992;37(7):338–40.

[95] Kyurkchan AG. On a method for solving the problem of wave diffraction by finite-size scatterers. Phys Dokl 1994;39(8):546–9.

[96] Kyurkchan AG. On solving the problem of plane wave scattering by a periodic interface of two media. Phys Dokl 1994;39(12):827–31.

[97] Kyurkchan AG, Kleev AI. Solution of wave diffraction problems on the scatterers of finite size by the pattern equations method. J Commun Technol Electron 1995;40 (6):897–905.

[98] Kyurkchan AG. On solving the problem of wave scattering by several bodies. Dokl Math 1996;53(3):453–61.

[99] Kyurkchan AG. Application of the method of pattern equations to the solution of multi-body scattering problem. J Commun Technol Electron 1996;41(1):33–8.

[100] Kyurkchan AG. A method for solving the problems of wave scattering at transparent obstacles. Dokl Math 1997;55(1):180–3.

[101] Kyurkchan AG, Manenkov SA. Wave scattering by a body immersed in a homogeneous half-space. Phys Dokl 1997;42(11):40–3.

[102] Kyurkchan AG, Manenkov SA. Scattering of waves by an inhomogeneity located close to the planar interface of two media. J Commun Technol Electron 1998;43(1):4–41.

[103] Kyurkchan AG, Manenkov SA. A new method for solving the problem of diffraction by a compact obstacle in a layered medium. Radiophys Quant Electron 1998;41 (7):589–99.

[104] Kyurkchan AG, Manenkov SA, Kleev AI, Sukov AI. Pattern equation method for solution of electromagnetic scattering problems. J Appl Electromagn 1999;2(1):17–31.

[105] Kyurkchan AG, Soloveichik AL. Scattering of waves by a periodic grating located close to a planar interface. J Commun Technol Electron 2000;45(4):357–64.

[106] Kyurkchan AG. Solution of vector scattering problem by the pattern equation method. J Commun Technol Electron 2000;45(9):970–5.

[107] Kalinchenko GA, Kyurkchan AG, Lerer AM, Manenkov SA, Soloveichik AL. Transmission and reflection factors of eigenmodes in a planar dielectric waveguide calculated

in the presence of nearby external objects. J Commun Technol Electron 2001;46 (9):1005–13.

[108]   Kyurkchan AG, Demin DB. Electromagnetic wave diffraction at impedance scatterers with piecewise-smooth boundaries. J Commun Technol Electron 2002;47(8):856–63.

[109]   Kyurkchan AG, Manenkov SA. Wave scattering at a group of bodies located in a planar stratified medium. J Commun Technol Electron 2002;47(11):1205–11.

[110]   Kyurkchan AG, Demin DB. Diffraction of electromagnetic waves by the dielectric scatterers. Sci Technol 2003;4(3):38–44 [in Russian].

[111]   Kyurkchan AG, Demin DB. Modeling of the wave scattering characteristics by bodies with a dielectric coating using impedance boundary conditions. Electromagn Waves Electron Syst 2003;8(11):68–78 [in Russian].

[112]   Kyurkchan AG, Demin DB. Simulation of wave scattering by bodies with an absorbing coating and by black bodies. Tech Phys 2004;49(2):165–73.

[113]   Kyurkchan AG, Demin DB. Pattern equation method for solving problems of diffraction of electromagnetic waves by axially symmetric dielectric scatterers. J Quant Spectrosc Radiat Transf 2004;89:237–55.

[114]   Kyurkchan AG, Demin DB. Electromagnetic wave diffraction at magnetodielectric and absorbing scatterers. J Commun Technol Electron 2004;49(11):1218–27.

[115]   Kyurkchan AG, Skorodumova EA. The solution of three-dimensional problems of wave diffraction at a group of bodies by the pattern equations method. Antenna 2005;94(3):55–9 [in Russian].

[116]   Kyurkchan AG, Skorodumova EA. Modeling the characteristics of the waves scattering by a group of scatterers. J Quant Spectrosc Radiat Transf 2006;100:207–19.

[117]   Kyurkchan AG, Skorodumova EA. Solution of the three-dimensional problem of wave diffraction by an ensemble of objects. Acoust Phys 2007;53(1):1–10.

[118]   Kyurkchan AG, Demin DB, Orlova NI. Use of the pattern equation method for solving the problems of scattering of electromagnetic waves by bodies covered with a dielectric. J Commun Technol Electron 2007;52(2):131–9.

[119]   Kyurkchan AG, Demin DB, Orlova NI. Solution based on the pattern equation method for scattering of electromagnetic waves by objects coated with dielectric materials. J Quant Spectrosc Radiat Transf 2007;106(1–3):192–202.

[120]   Kyurkchan AG, Skorodumova EA. The problem of electromagnetic wave diffraction at a group of bodies solved via the pattern equation method. J Commun Technol Electron 2007;52(11):1229–36.

[121]   Kyurkchan AG, Skorodumova EA. Solving the diffraction problem of electromagnetic waves on objects with a complex geometry by the pattern equation method. J Quant Spectrosc Radiat Transf 2008;109(8):1417–29.

[122]   Kyurkchan AG, Demin DB. Solution to the problem of scattering of electromagnetic waves by layered inhomogeneous bodies. J Commun Technol Electron 2009;54 (1):52–62.

[123]   Kyurkchan AG, Skorodumova EA. Solving the problems of electromagnetic waves scattering by complex-shaped dielectric bodies via pattern equations method. J Quant Spectrosc Radiat Transf 2009;110(14–16):1335–44.

[124]   Kyurkchan AG, Demin DB. Solution of problem of electromagnetic waves scattering on inhomogeneously layered scatterers using pattern equations method. J Quant Spectrosc Radiat Transf 2009;110(14–16):1345–55.

[125]   Kyurkchan AG, Skorodumova EA. Solution of the problem of electromagnetic wave diffraction by complex dielectric bodies by means of the pattern equation method. J Commun Technol Electron 2009;54(6):625–36.

[126]   Kyurkchan AG, Smirnova NI. T-matrix and pattern equations methods for solution of diffraction problems. Electromagn Waves Electron Syst 2009;15(8):27–35 [in Russian].

[127] Kyurkchan AG, Smirnova NI. Pattern equation method as an alternative to T-matrix method, In: Proceedings of the International Conference "Days on Diffraction 2010". June 8–11; 2010. p. 113–9.

[128] Kyurkchan AG, Demin DB. Solution of problems of electromagnetic wave scattering by bodies with an anisotropic impedance with the help of the pattern equation method. J Commun Technol Electron 2011;56(2):134–41.

[129] Kyurkchan AG, Demin DB. Pattern equation method for the solution of electromagnetic scattering by axially-symmetric particles with complex anisotropic surface impedance. Rev Mate Teor Aplic 2013;20(1):1–20.

[130] Kyurkchan AG, Smirnova NI. About one new method of solving diffraction problems, In: Proceedings of the International Conference "Days on Diffraction 2011". May 30–June 3; 2011. p. 117–21.

[131] Kyurkchan AG, Smirnova NI. The solution of diffraction problems by the method of elementary scatterers. Electromagn Waves Electron Syst 2011;16(8):5–10 [in Russian].

[132] Ufimtsev PYa. Method of edge waves in the physical theory of diffraction. Moscow: Sov. Radio; 1961 [in Russian].

[133] Kyurkchan AG, Manenkov SA. Application of various orthogonal coordinate frames for simulating the wave scattering by a group of bodies. J Commun Technol Electron 2012;57(9):1001–8.

[134] Kyurkchan AG, Smirnova NI. Solving diffraction problems on compact scatterers by hybrid approach using continued boundary conditions method, In: Abstracts of International conference "Days on diffraction 2012" Saint Petersburg, May 28–June 1; 2012. p. 102–3.

[135] Bogomolov GD, Kleev AI. Calculation of open resonators with the help of a modified method of continued boundary conditions. J Commun Technol Electron 2011;56 (10):1179–85.

[136] Kyurkchan AG, Smirnova NI. On modification of the continued boundary conditions method, In: Proceedings of the International Conference "Mathematical Methods in Electromagnetic Theory". Kharkiv, Ukraine, August 28–30; 2012. p. 431–4.

[137] Fedoryuk MV. Asymptotics. Integrals and series. Moscow: Nauka; 1987 [in Russian].

[138] Kantorovich LV, Akilov GP. Functional analysis. Moscow: Nauka; 1977 [in Russian].

[139] Twersky V. Multiple scattering of electromagnetic waves by arbitrary configurations. J Math Phys 1967;8(3):589–605.

[140] Titchmarsh EC. The zeta-function of Riemann. Oxford: Oxford Press; 1930.

[141] Kyurkchan AG, Manenkov SA, Smirnov VI. Analysis of diffraction of a plane wave and of the field of a point source by a multirow grating. J Commun Technol Electron 2011;56(9):1057–68.

[142] Konyukhova NB, Pak TV. Diffraction of a plane acoustic wave by a hard prolate spheroid. Moscow: Computing Center AN SSSR; 1985 [in Russian].

[143] Haus HA. Waves and fields in optoelectronics. Englewoods Cliffs, New Jersey: Prentice-Hall; 1984.

[144] Kyurkchan AG. On the realizability of directionality diagrams of antennas created by currents distributed on a closed curve. Sov Phys Dokl 1982;27(7):577–9.

[145] Kyurkchan AG. Inverse scattering problem for the Helmholtz equation. Sov Phys Dokl 1984;29(3):48–51.

[146] Kyurkchan AG. Analytic properties of patterns in the complex plane. In: Katsenelenbaum BZ, Sivov AN, editors. Electrodynamics of antennas with semi-transparent surfaces: methods of constructive synthesis. Moscow: Nauka; 1989.

[147] Kyurkchan AG. On the minimal domain occupied by currents implementing a given directional pattern. J Commun Technol Electron 2000;45(3):258–60.

INDEX

Note: Page numbers followed by *f* indicate figures, and *t* indicate tables.

Printed in the United States
By Bookmasters